Orchard Hideout

Circles and Coordinate Geometry

Teacher's Guide

This material is based upon work supported by the National Science Foundation under award numbers ESI-9255262, ESI-0137805, and ESI-0627821. Any opinions, findings, and conclusions or recommendations expressed in this publication are those of the authors and do not necessarily reflect the views of the National Science Foundation.

Key Curriculum
1150 65th Street
Emeryville, California 94608
email: editorial@keypress.com
www.keycurriculum.com

First Edition Authors

Dan Fendel, Diane Resek, Lynne Alper, and Sherry Fraser

Contributors to the Second Edition

Sherry Fraser, Jean Klanica, Brian Lawler, Eric Robinson, Lew Romagnano, Rick Marks, Dan Brutlag, Alan Olds, Mike Bryant, Jeri P. Philbrick, Lori Green, Matt Bremer, Margaret DeArmond

Project Editors

Sharon Taylor, Josephine Noah

Consulting Editor

Mali Apple

Editorial Assistants

Juliana Tringali, Emily Reed

Professional Reviewer

Rick Marks, Sonoma State University

Calculator Materials Editor

Josephine Noah

Math Checker

Carrie Gongaware

Production Editor

Andrew Jones

Production Director

Christine Osborne

Executive Editor

Josephine Noah

Mathematics Product Manager

Elizabeth DeCarli

Publisher

Steven Rasmussen

Contents

Introduction

Activity Notes

Introduction

Orchard Hideout Unit Overview

Intent

The activities of this unit develop major concepts of coordinate geometry and circles, in the context of a complex central unit problem.

Mathematics

The central unit problem concerns a couple who have planted an orchard of trees in careful rows and columns on a circular lot. The couple realize that, after a while, the trunks of their trees will become so thick that they will no longer be able to see out from the center of the orchard. In other words, the orchard will become a "hideout." The main unit question is this:

How soon after the couple plant the orchard will the center of the lot become a true "orchard hideout"?

Students' search for the answer to this question leads them to the study of several aspects of geometry.

Students use the Pythagorean theorem to measure distances within the orchard, leading to development of the distance formula. As a sidelight to their work with the distance formula, students construct the general equation of a circle.

Giving the initial size of the trees in terms of circumference and the growth rate in terms of cross-sectional area motivates development of the area and circumference formulas for a circle.

While solving the unit problem, students encounter a variety of tangents (both figuratively and literally). One result is a proof that a tangent to a circle is perpendicular to the radius at the point of tangency. They use the technique of completing the square to put certain quadratic equations into standard form to find the radius and center of the circles they represent. Other ideas arise through the unit's POWs. For example, students prove basic facts about perpendicular bisectors and angle bisectors, thereby establishing the existence of both circumscribed and inscribed circles for triangles.

The main concepts and skills students will encounter and practice during the unit are summarized below.

Coordinate geometry

- Using the Cartesian coordinate system to organize a complex problem

- Developing and applying the distance formula

- Developing the standard form for the equation of a circle with a given center and radius

- Finding the distance from a point to a line in a coordinate setting
- Developing and applying the midpoint formula

Circles

- Using similarity to see that the circumference of a circle should be a constant times its radius, and that the area of a circle should be a constant times the square of its radius
- Finding formulas for the perimeter and area of regular polygons circumscribed about a circle
- Using circumscribed polygons to see that the "circumference coefficient" for the circle is twice the "area coefficient" for the circle
- Defining π and understanding why it appears in the formulas for both the circumference and the area of a circle
- Developing and applying formulas for the circumference and area of a circle

Synthetic geometry

- Identifying and describing a set of points satisfying a geometric condition
- Discovering and proving that the set of points equidistant from two given points is the perpendicular bisector of the segment connecting the given points
- Defining the distance from a point to a line and proving that the perpendicular distance is the shortest
- Discovering and proving that any line through the midpoint of a segment is equidistant from the endpoints of the segment
- Discovering and proving that the set of points equidistant from two intersecting lines consists of the bisectors of the angles formed by the lines

Algebra

- Using the technique of completing the square to transform equations of circles into standard form
- Using algebra in a variety of proofs involving coordinates and angles

Logic

- Understanding and using the phrases "if-then" and "if and only if" in definitions and proofs
- Working with converses

Miscellaneous

- Using symmetry to help analyze a problem
- Learning about Pythagorean triples

Progression

In *Orchards and Mini-Orchards*, students explore simplified versions of the unit problem and begin to create a general plan for solving it.

Early in the unit, students see the usefulness of representing the orchard using the coordinate system. They recognize that there is a tree at every lattice point within the lot (these are the points with integer coordinates) except the origin. In *Coordinates and Distance*, students' work with coordinate geometry builds on the Pythagorean theorem and includes development of the distance formula and the midpoint formula. They then use the distance formula to construct the equation of a circle with center at the origin.

Another key component of the unit is finding the distance from a given lattice point to a given line of sight out of the orchard. This is addressed in *Equidistant Points and Lines*. To do this, students develop and combine a variety of ideas from coordinate geometry, synthetic geometry, and trigonometry.

An important idea from synthetic geometry is the principle that a line through the midpoint of a line segment is equidistant from the segment's endpoints. (Students develop and prove this principle through an activity that grows out of one of the unit's POWs.) The actual computation of distance from a point to a line is done in several ways, using both similarity and right-triangle trigonometry, thus building on students' work in Years 1 and 2.

To apply information about the growth rate of the trees, students need to understand how the circumference, area, and radius of a circle are related. They begin in *All About Circles* by comparing circles to circumscribed squares and then use principles of similarity to see two key facts.

- The circumference of a circle is a fixed number (the "circumference coefficient") times its radius.

- The area of a circle is a fixed number (the "area coefficient") times the square of its radius.

By looking at various other circumscribed regular polygons, students gain further insight into the formulas for the area and circumference of circles and the connection between the two. In particular, they see, through both polygon examples and an area analysis, that the circumference coefficient is exactly twice the area coefficient. This observation leads to the definition of π and to the formulas $C = 2\pi r$ and $A = \pi r^2$.

In *Cable Complications*, students work on a complex activity concerning distance from a point to a line in the context of their orchard, requiring that they assemble many of the pieces from earlier assignments and apply them toward the unit problem solution. They also work further with the equation of a circle, expanding it to a more general form that accommodates arbitrary coordinates for the center and looking at how to transform an equation into this form through completing the square.

Toward the end of the unit, students determine which line of sight will remain unblocked the longest. Their search for this last line of sight begins in *Lines of Sight* with a detailed study of distances to lines of sight in an orchard of radius 3. They see that the last unblocked lines of sight are those closest to the coordinate axes.

They then "eyeball" lines of sight for an orchard of radius 6, and their intuition suggests that this principle regarding the location of the last line of sight should be true in general. They learn that this intuition is correct but that the proof is quite advanced and requires concepts they have not yet studied.

The unit ends with students working in groups to develop a complete solution to the central unit problem, finding both the distance to the last line of sight from its nearest lattice point and the time required for the couple's trees to reach the necessary size.

Orchards and Mini-Orchards

Coordinates and Distance

Equidistant Points and Lines

All About Circles

Cable Complications

Lines of Sight

Pacing Guides

50-Minute Pacing Guide (26 days)

Day	Activity	In-Class Time Estimate
1	Orchards and Mini-Orchards	
	Orchard Hideout	30
	Reference: A Geometric Summary	15
	Homework: Geometry and a Mini-Orchard	5
2	Discussion: Geometry and a Mini-Orchard	15
	Presentations: A Geometric Summary	30
	Introduce: POW 1: Equally Wet	5
	The Standard POW Write-up	0
	Homework: Only Two Flowers	0
3	Discussion: Only Two Flowers	10
	A Perpendicularity Proof	40
	Homework: From Two Flowers to Three	0
4	Discussion: From Two Flowers to Three	10
	More Mini-Orchards	35
	Homework: In, On, or Out?	5
5	Coordinates and Distance	
	Discussion: In, On, or Out?	50
	Homework: Other Trees	0
6	Discussion: Other Trees	10
	Sprinkler in the Orchard	40
	Homework: The Distance Formula	0
7	Discussion: The Distance Formula	15
	How Does Your Orchard Grow?	35
	Homework: A Snack in the Middle	0
8	Discussion: A Snack in the Middle	15
	Presentations: POW 1: Equally Wet	35

	Equidistant Points and Lines	
	Homework: *Proving with Distance–Part I*	0
9	Discussion: *Proving with Distance–Part I*	10
	Down the Garden Path	40
	Homework: *Perpendicular and Vertical*	0
10	Discussion: *Perpendicular and Vertical*	15
	Discussion: *Down the Garden Path*	25
	Introduce: *POW 2: On Patrol*	10
	Homework: *Proving with Distance–Part II*	0
11	Discussion: *Proving with Distance–Part II*	15
	All About Circles	
	Squaring the Circle	35
	Homework: *Using the Squared Circle*	0
12	Discussion: *Using the Squared Circle*	15
	Hexagoning the Circle	35
	Homework: *Octagoning the Circle*	0
13	Discussion: *Octagoning the Circle*	15
	Polygoning the Circle	35
	Homework: *Another Kind of Bisector*	0
14	Discussion: *Another Kind of Bisector*	10
	Discussion: *Polygoning the Circle*	15
	Optional: *The Story of Pi* video	25
	Homework: *Proving Triples*	0
15	Discussion: *Proving Triples*	10
	Presentations: *POW 2: On Patrol*	25
	Introduce: *POW 3: A Marching Strip*	10
	Homework: *Orchard Growth Revisited*	5
16	Discussion: *Orchard Growth Revisited*	10
	Cable Complications	
	Cable Ready	40
	Homework: *Going Around in Circles*	0
17	Discussion: *Going Around in Circles*	10
	Cable Ready (continued)	40

	Homework: *Daphne's Dance Floor*	0
18	Discussion: *Cable Ready*	35
	Discussion: *Daphne's Dance Floor*	15
	Homework: *Defining Circles*	0
19	Discussion: *Defining Circles*	15
	The Standard Equation of the Circle	20
	Homework: *Completing the Square and Getting a Circle*	15
20	Discussion: *Completing the Square and Getting a Circle*	10
	Lines of Sight	10
	The Other Gap	30
	Homework: *Cylindrical Soda*	0
21	Discussion: *Cylindrical Soda*	10
	Discussion: *The Other Gap*	10
	Lines of Sight for Radius Six	30
	Homework: *Orchard Time for Radius Three*	0
22	Discussion: *Orchard Time for Radius Three*	10
	Hiding in the Orchard	40
	Homework: *Big Earth, Little Earth*	0
23	Discussion: *Big Earth, Little Earth*	15
	Presentations: *Hiding in the Orchard*	35
	Homework: *Beginning Portfolios*	0
24	Discussion: *Beginning Portfolios*	15
	Presentations: *POW 3: A Marching Strip*	25
	Homework: *"Orchard Hideout" Portfolio*	10
25	*In-Class Assessment*	50
	Homework: *Take-Home Assessment*	0
26	Discussion: *"Orchard Hideout" Portfolio* and *Unit Reflection*	10
	Exam Discussion	40

90-Minute Pacing Guide (17 days)

Day	Activity	In-Class Time Estimate
1	*Orchards and Mini-Orchards*	
	Orchard Hideout	30
	Reference: *A Geometric Summary*	10
	Geometry and a Mini-Orchard	40
	POW 1: Equally Wet	10
	The Standard POW Write-up	0
	Homework: *Only Two Flowers*	0
2	Discussion: *Only Two Flowers*	10
	Presentations: *A Geometric Summary*	40
	A Perpendicularity Proof	40
	Homework: *From Two Flowers to Three*	0
3	Discussion: *From Two Flowers to Three*	10
	More Mini-Orchards	40
	Coordinates and Distance	
	Other Trees	35
	Homework: *In, On, or Out?*	5
4	Discussion: *In, On, or Out?*	50
	Sprinkler in the Orchard	40
	Homework: *The Distance Formula*	0
5	Discussion: *The Distance Formula*	15
	How Does Your Orchard Grow?	35
	A Snack in the Middle	40
	Equidistant Points and Lines	
	Homework: *Proving with Distance–Part I*	0
6	Discussion: *Proving with Distance–Part I*	10
	Presentations: *POW 1: Equally Wet*	35
	Down the Garden Path	45

	Homework: *Perpendicular and Vertical*	0
7	Discussion: *Perpendicular and Vertical*	10
	Discussion: *Down the Garden Path*	20
	Introduce: *POW 2: On Patrol*	5
	All About Circles	
	Squaring the Circle	30
	Using the Squared Circle	25
	Homework: *Proving with Distance–Part II*	0
8	Discussion: *Proving with Distance–Part II*	10
	Discussion: *Using the Squared Circle*	10
	Hexagoning the Circle	35
	Another Kind of Bisector	35
	Homework: *Octagoning the Circle*	0
9	Discussion: *Octagoning the Circle*	15
	Polygoning the Circle	50
	Optional: *The Story of Pi* video	25
	Homework: *Proving Triples*	0
10	Discussion: *Proving Triples*	15
	Orchard Growth Revisited	40
	Cable Complications	
	Cable Ready	35
	Homework: *Going Around in Circles*	0
11	Discussion: *Going Around in Circles*	15
	Discussion: *Cable Ready*	75
	Homework: *Daphne's Dance Floor*	0
12	Discussion: *Daphne's Dance Floor*	15
	Presentations: *POW 2: On Patrol*	25
	Introduce: *POW 3: A Marching Strip*	10
	Defining Circles	40
	Homework: *Cylindrical Soda*	0
13	Discussion: *Cylindrical Soda*	10
	The Standard Equation of the Circle	20
	Lines of Sight	10

	The Other Gap	35
	Homework: *Completing the Square and Getting a Circle*	15
14	Discussion: *Completing the Square and Getting a Circle*	10
	Lines of Sight for Radius Six	30
	Orchard Time for Radius Three	40
	Homework: *Big Earth, Little Earth*	10
15	Discussion: *Big Earth, Little Earth*	10
	Hiding in the Orchard	80
	Homework: *Take-Home Assessment*	0
16	*In-Class Assessment*	60
	Homework: *Beginning Portfolios*	30
17	Presentations: *POW 3: A Marching Strip*	25
	Exam Discussion	40
	Discussion: *Beginning Portfolios*	10
	Unit Reflection	15
	Homework: *"Orchard Hideout" Portfolio*	0

Materials and Supplies

All IMP classrooms should have a set of standard supplies, described in the section "Materials and Supplies for the IMP Classroom" in *A Guide to IMP.* You'll also find a comprehensive list of materials needed for all Year 3 units in the section "Materials and Supplies for Year 3" in the *Year 3 Teacher's Guide* general resources.

Listed here are the supplies needed for this unit. Also available are general and activity-specific blackline masters, for transparencies or for student worksheets, in the "Blackline Masters" section in *Orchard Hideout* Unit Resources.

Orchard Hideout Materials

- Paper cups or coins (at least 30 per group)
- Spaghetti or string (one piece per group)
- Large-grid poster paper
- Poster materials
- String or compass (one for each pair of students)
- (Optional) *The Story of Pi* video

More About Supplies

Graph paper is a standard supply for IMP classrooms. Blackline masters of 1-Centimeter Graph Paper, ¼-Inch Graph Paper, and 1-Inch Graph Paper are provided, for you to make copies and transparencies.

Assessing Progress

Orchard Hideout concludes with two formal unit assessments. In addition, there are many opportunities for more informal, ongoing assessments throughout the unit. For more information about assessment and grading, including general information about the end-of-unit assessments and how to use them, consult the *Year 3: A Guide to IMP* resource.

End-of-Unit Assessments

This unit concludes with in-class and take-home assessments. The in-class assessment is intentionally short so that time pressures will not affect student performance. Students may use graphing calculators and their notes from previous work when they take the assessments. You can download unit assessments from the *Orchard Hideout* Unit Resources.

Ongoing Assessment

One of the primary tasks of the classroom teacher is to assess student learning. Although the assigning of course grades may be part of this process, assessment more broadly includes the daily work of determining how well students understand key ideas and what level of achievement they have attained on key skills, in order to provide the best possible ongoing instructional program for them.

Students' written and oral work provides many opportunities for teachers to gather this information. We make some recommendations here of written assignments and oral presentations to monitor especially carefully that will give you insight into student progress.

- *Sprinkler in the Orchard*
- *Proving with Distance—Part I* or *Proving with Distance—Part II*
- *Polygoning the Circle*
- *Orchard Growth Revisited*
- *Cable Ready*
- *Hiding in the Orchard*

Discussion of Unit Assessments

Have students volunteer to explain their work on each of the problems. Encourage questions and alternate explanations from other students.

In-class Assessment

You can have volunteers present each problem. They should realize that in Question 1, Madie and Clyde want to put the transmitter at the midpoint of the segment connecting their positions, which is 390 feet from each of them. Questions 2 and 3 are straightforward applications of the equation of a circle.

Take-home Assessment

Question 1 involves the formulas for the circumference and area of a circle as well as the principles for volume and lateral surface area of a cylinder. In Question 1a, students have to remember to divide by 100 to account for the fact that a quart of sealant covers 100 square feet of surface. Students might leave answers in terms of π, but it probably makes more sense in the context for them to give decimal approximations instead.

Here are the answers for Question 1.

- Area = $\pi \; 5^2 \approx 78.5$ square feet

- Volume = $3.5 \cdot$ area ≈ 275 cubic feet

- Lateral surface area = $3.5 \cdot 2\pi \cdot 5 \approx 110$ square feet, so Madie needs about 1.1 quarts of sealant. Thus, she should purchase 2 quarts to make sure she has enough.

The wording of Question 2a means that the tree where the target will go must have its center 5 units from (–4, –15), but that tree cannot have the same *x*-coordinate or *y*-coordinate as (–4, –15). [For example, Madie and Clyde could place the target on the tree at (–7, –11), which is 3 units "down" and 4 units "over" from

(–4, –15).]

Question 2b involves a fairly straightforward application of the midpoint formula.

Supplemental Activities

Each unit contains a variety of activities that you can use to supplement the regular unit material. These activities are included at the end of the student pages for the unit and fall roughly into two categories.

Reinforcements increase students' understanding of and comfort with concepts, techniques, and methods that are discussed in class and are central to the unit.

Extensions allow students to explore ideas beyond those presented in the unit, including generalizations or abstractions of ideas that are part of the main unit.

The supplemental activities are presented in the teacher's guide and in the student book in the approximate sequence in which you might use them. Listed here are specific recommendations about how each activity might work within the unit. You may wish to use some of these activities, especially the later ones, after the unit is completed.

Teaching suggestion: This appendix contains several activities that are quite abstract. These are included for students who are interested in pursuing the geometric ideas, but we suggest that you exercise caution in assigning them to the whole class.

***Right and Isosceles* (reinforcement)** This activity makes a natural follow-up to the discussion following *Geometry and a Mini-Orchard* about the length of the diagonal of the unit square.

***The Perpendicular Bisector Converse* (extension)** The statement to be proved in this activity is the converse of a statement that students prove in *A Perpendicularity Proof*. The activity provides the necessary diagram, so this is a good starting place for students to work on proofs of this type on their own.

***Counting Trees* (extension)** This activity builds on the discussion of symmetry in *Other Trees*. Question 3 is related to the supplemental problem *Right and Isosceles*. The goal is for students to see that every orchard with a positive integer radius has four boundary trees on the axes and that the number of boundary trees not on the axes is always a multiple of 8.

***Perpendicular Bisectors by Algebra* (extension)** In this activity, students apply the distance formula (developed in *The Distance Formula*) to get the equation of the perpendicular bisector of a specific line segment and then generalize to an arbitrary segment.

***Midpoint Proof* (extension)** In *A Snack in the Middle*, students are asked to prove only one part of the assertion that (27, 10) is the midpoint of the segment connecting (24, 6) and (30, 14). In this activity, students are asked to complete the proof.

Although the activity suggests doing this by finding a linear equation through all three points, some students may provide a more geometric argument, perhaps based on an intuitive notion of slope.

***Why Do They Always Meet?* (extension or reinforcement)** This activity continues ideas from *POW 1: Equally Wet*, and you can assign it any time after the POW presentations. Several other supplemental problems refer to this one, so you may want to discuss it in class.

***Inscribed Angles* (extension)** In this activity, students are asked to combine algebra and geometry to prove that an angle inscribed in a semicircle must be a right angle. Because the activity builds on the concept of a circle circumscribed about a triangle, you should assign it after the discussion of *POW 1: Equally Wet*.

***More Inscribed Angles* (extension)** This activity builds directly on the supplemental problem *Inscribed Angles*, and so students should probably not attempt *More Inscribed Angles* unless they have done the preceding supplemental problem.

***Angles In and Out* (extension and reinforcement)** Although this activity refers to the supplemental problems *Inscribed Angles* and *More Inscribed Angles*, it is actually both independent of and simpler than them. It is essentially an application of the fact that the sum of the angles of any triangle is 180°.

This activity can be assigned at any time, but it is placed after the problems on inscribed angles because that topic provides a good context for the activity. It is labeled both an extension and a reinforcement because it is fairly elementary yet it introduces a principle that is not part of the main unit.

***Midpoint Quadrilaterals* (extension)** This follow-up to *Proving with Distance—Part I* provides a good application of the distance formula and of principles about similarity and parallel lines.

***Equidistant Lines* (extension)** The proof outlined in this activity is a follow-up to *Down the Garden Path*, and the problem can be assigned anytime after the discussion of that activity.

***Right in the Center* (reinforcement)** Questions 1 and 2 of this activity are similar to problems in *Proving with Distance—Part I* and *Proving with Distance—Part II*, and can be used after those homework assignments.

Question 3, however, involves the general equation for a circle, which is not discussed until *Defining Circles*, so you may wish to wait until that point in the unit before assigning this supplemental problem.

The supplemental problem *Hypotenuse Median* generalizes the example in this activity.

***Thirty-Sixty-Ninety* (reinforcement)** Students' work with hexagons in *Hexagoning the Circle* provides a good context and motivation for this activity.

***Darts* (reinforcement)** This problem provides a simple application of the area formula developed through the activities culminating in *Polygoning the Circle*. However, because the problem is phrased in terms of expected value, students may need some review of concepts about probability.

***More About Triples* (extension)** The topic of Pythagorean triples, though inspired by the geometry of the right triangle, is mostly about number theory. Interested students can use the two questions in this activity as a starting point for a more extensive report. You can assign this activity anytime after *Proving Triples.*

***The Inscribed Circle* (extension)** This activity is a follow-up to the discussion of *POW 2: On Patrol* and is analogous to the supplemental problem *Why Do They Always Meet?*

***Medians and Altitudes* (extension)** The two problems in this activity develop two principles similar to those in the main unit—that the perpendicular bisectors of the sides of a triangle meet in a single point and that the bisectors of the angles of a triangle meet in a single point. The analogous statements for medians and altitudes are considerably more difficult to prove, however, and you may want to suggest that students consult a traditional geometry textbook for ideas.

You can use this problem after students have seen the two simpler principles for perpendicular bisectors and angle bisectors in *POW 1: Equally Wet* and *POW 2: On Patrol.*

***Hypotenuse Median* (extension)** This activity generalizes the example in the supplemental problem *Right in the Center*. The activity suggests two different approaches to the proof, so it is a good assignment to synthesize ideas from the unit.

***Not Quite a Circle* (extension)** This exploration is a natural follow-up to students' work with the equation of a circle. Although students will have the necessary background after *In, On, or Out?*, you may want to wait until later in the unit before assigning this problem, because it is quite open-ended.

***Knitting* (extension)** This supplemental problem involves the principle that the volume of a sphere is proportional to the cube of the radius. This principle is not discussed in the unit, but *Daphne's Dance Floor* explores similar ideas for circumference and area of a circle, and *Cylindrical Soda* involves related ideas about the volume of a cylinder.

***What's a Parabola?, Creating Parabolas, Coordinate Ellipses and Hyperbolas, Another View of Ellipses and Hyperbolas, Ellipses and Hyperbolas by Points and Algebra, Generalizing the Ellipse,* and *Moving the Ellipse* (extensions)** These problems form something like a mini-course on conic sections, primarily from a coordinate perspective. They are a natural extension to the work in this unit on coordinate geometry and the equation of a circle, and might be assigned after the work on the equation of a circle in *The Standard Equation of the Circle.*

Orchards and Mini-Orchards

Intent

The activities in *Orchards and Mini-Orchards* introduce students to the unit problem and to the first POW. Students also work with a smaller orchard to develop the equation of a circle with the center at the origin.

Mathematics

A Geometric Summary provides students with a structured review of the geometric principles that they have studied in previous courses. These principles are critical to developing the understanding they will need in order to solve the unit problem. In preliminary work with a simplified version of the orchard, students calculate distances using the Pythagorean theorem. This lays the groundwork for developing the equation of a circle and, later, the distance formula. Consideration of lines of sight in the orchard also brings up the issue of tangency to a circle.

Several assignments guide students through preliminary work on POW 1. In the process, students discover and then prove that a point is equidistant from the endpoints of a line segment if and only if it lies on the perpendicular bisector of the segment. This provides a context for introducing the converse and "if-then" and "if and only if" statements. Optionally, this can also be used to introduce geometric constructions. The final solution to *POW 1: Equally Wet* introduces the concept of the circumcenter of a polygon (if the polygon can be circumscribed), and the fact that, if the polygon can be circumscribed, the circumcenter lies at the intersection of the perpendicular bisectors of the sides of the polygon.

Progression

Two separate threads are interwoven: preliminary investigation of the unit problem introduced in *Orchard Hideout* and initial work on *POW 1: Equally Wet*.

At this stage, work on the unit problem focuses primarily on visualizing the situation. After an initial review of geometric concepts from previous courses in *A Geometric Summary*, *Geometry and a Mini-Orchard* exposes students to the idea that there is a line of sight from the center of the orchard that will be the last one blocked as the tree radii increase, which will be referred to throughout the unit as the *last line of sight*. In *More Mini-Orchards*, students discover that it is not necessary for the trees to actually touch one another in order to block the line of sight, and a line of sight that is just touching a tree is described as *tangent* to the circle that represents the perimeter of the trunk. *In, On, or Out?* asks students to determine whether trees with particular coordinates lie within, on, or outside of a circle with radius 10. Students will use the Pythagorean theorem to address trees that are in doubt and to develop the equation of a circle centered at the origin.

POW 1: Equally Wet asks students to explain how to determine the location of a sprinkler so that it will be equidistant from any given number of flowers whose locations are known. *Only Two Flowers* simplifies that problem to a case with only two flowers, leading students to discover that the sprinkler must be placed on the perpendicular bisector of the line segment joining the two flowers, and *A Perpendicularity Proof* asks students to prove this. *From Two Flowers to Three* challenges students to find a suitable sprinkler placement for a particular arrangement of three flowers, setting the stage for discussion of the circumcenter of a polygon upon later completion of work on the POW.

Orchard Hideout
A Geometric Summary
Geometry and a Mini-Orchard
POW 1: Equally Wet
Only Two Flowers
A Perpendicularity Proof
From Two Flowers to Three
More Mini-Orchards
In, On, or Out?

Orchard Hideout

Intent

Students are introduced to the situation in the unit problem, identifying the key questions that will need to be answered.

Mathematics

Making models will help students to clearly visualize the situation. Students will identify issues to be clarified, assumptions to be made, and key questions to be answered.

Progression

Students begin by making a physical model of a similar but smaller orchard. They are then asked to make a list of questions they will need to ask. Subsequent discussion will begin with ensuring that everyone understands the unit problem situation. Discussion of the students' questions will lead to three simplifying assumptions: the tree trunks are all identical and cylindrical; the trees all grow at the same rate; and a tree can be planted if the center of its trunk would fall on or within the circle. Finally, two key questions necessary to the solution should be identified: how big do the tree trunks have to become, and how long will that take?

Approximate Time

20 minutes for activity
10 minutes for discussion

Classroom Organization

Small groups, followed by whole-class discussion

Materials

Paper cups or coins (to represent the trees)
Spaghetti or string (to represent the lines of sight)
Poster grid paper
Graph paper

Doing the Activity

Have students read the unit's title activity, *Orchard Hideout,* and work on it in their groups. They will encounter the main question of the unit, "How soon after Madie and Clyde plant their orchard will the center of the lot become a true 'orchard hideout'?" before they begin.

Question 1 asks groups to make a model of the situation. You should provide materials for students to do so, such as paper cups or coins for the trees and spaghetti or string for the lines of sight.

Because there may not be enough materials, time, or space for students to build a complete model with 50 trees in each direction from the center, assure them that it is okay to make a model with a smaller radius. Emphasize that the intent of Question 1 is for them to get a clear idea of what the situation looks like, and not for them to use the model to solve the problem.

As you circulate around the classroom, encourage students to experiment with trees and orchards of different sizes, and verify in each case whether it's possible to see out from the center of the orchard.

Students may also want to make a sketch that shows the situation more schematically, so you should have graph paper available for that.

Question 2 asks students to make a list of questions they need to ask in order to understand the problem better. They should try to answer some of their own questions, and save the rest for the whole-class discussion.

Discussing and Debriefing the Activity

After groups have had a chance to wrestle with the problem, ask one member of each group to share the group's model with the class.

Be sure that students can create a general three-dimensional model of what the orchard looks like before moving to two-dimensional sketches. Ask, What is a true "orchard hideout"? The orchard will be a true hideout if every possible line drawn from the center of the orchard hits the trunk of some tree in the orchard.

Use the term *line of sight* to refer to any line along which Madie and Clyde might look outward from the center of the orchard (whether or not the line is blocked by a tree). Bring out that as the trees grow, more lines of sight become blocked.

A Sketch of a Smaller Orchard
If no group used a two-dimensional sketch to describe the situation, have students draw a sketch for a small orchard. For instance, if the orchard had a radius of 6 units (instead of 50), the arrangement of trees would look like the following diagram, in which each dot represents a tree. (Don't get into details now about exactly where each row ends. Students will address that issue in *In, On, or Out?*)

During the discussion, be sure that students use the terms *circle, center*, and *radius* correctly. You may need to tell them that the plural of *radius* is *radii*. (In the discussion following *In, On, or Out?*, students will develop a formal definition of the term *circle*. For now, an intuitive approach will suffice.)

Open Questions
Have groups share the questions they came up with and the answers they may have found, and give other students a chance to answer the ones the groups could not.

Discuss the need to make some preliminary simplifying assumptions to obtain a useful mathematical model for this problem. Get the class to agree on these assumptions.
- The trees are all identical, and each has a cylindrical trunk. (We are interested only in the trees' trunks, and not in other aspects of their growth, like their branches or fruit.)
- The trees all grow at the same rate. (Students will have to know this rate in order to answer the "how soon" part of the main unit question.)
- A tree can be planted as long as its proposed center is on or within the boundary, even if the tree will eventually grow beyond the boundary.

Students may want to know what the actual distance is between centers of trees in a given row. Tell them that for now, they will simply use this distance as our unit length.

Two Main Questions
The main unit question can be thought of in terms of these two more specific questions.
- How big do the tree trunks have to become for the center of the orchard to be a hideout?

- How long will it take for the tree trunks to reach that size?

Post these questions (or similar ones) so students can refer to them over the course of the unit.

Comment: The first question, though difficult, is a purely geometric problem. To answer the second question, students will need to know something about how fast the trees grow, how big they are initially, and so on. You can tell them that they will get this sort of information later in the unit.

Key Question

What is a true "orchard hideout"?

How big do the tree trunks have to become for the center of the orchard to be a hideout?

How long will it take for the tree trunks to reach that size?

A Geometric Summary

Intent

Students review principles of geometry and trigonometry that they studied in previous courses.

Mathematics

A Geometric Summary contains a synopsis of basic definitions and principles about geometry (including trigonometry). Those topics include polygon angle sums, similarity, congruence, the Pythagorean theorem, trigonometric functions, parallel lines, perimeter, area, volume, and surface area.

Progression

Students are presented here with a summary of key geometric concepts for review and reference. Working in pairs or small groups, students will prepare and deliver a brief presentation on an assigned topic.

Approximate Time

15 minutes to assign topics
30 to 45 minutes to prepare presentations (at home or in class)
30 to 40 minutes for presentations

Classroom Organization

Pairs or small groups

Materials

Poster materials

Doing the Activity

Reading the entirety of *A Geometric Summary* is Part I of *Geometry and a Mini-Orchard*. We suggest that you ask students to prepare presentations on specific portions of this summary. The principles are stated rather concisely here, and student presentations should include more details—especially diagrams—to explain the meaning of the terms and ideas involved. You might have students work in pairs or larger groups on these presentations, perhaps allowing class time for them to do so.

Discussing and Debriefing the Activity

Have students make the presentations. As each topic is presented, be sure to allow time for questions.

You may want to spread these presentations out over several days. However, students will need to use ideas about either congruence or the Pythagorean theorem in connection with *A Perpendicularity Proof*, so it will be helpful if that part of the material is discussed before the class begins that activity.

Geometry and a Mini-Orchard

Intent

Students are introduced to the concept of a last line of sight in the unit problem situation.

Mathematics

Students will use the Pythagorean theorem, trigonometry, or area calculations to calculate the radii of two diagonally adjacent trees on the orchard grid that are touching.

Progression

Part I of this activity asks students to review the reference material in *A Geometric Summary*. Part II simplifies the unit problem situation to an orchard of radius 1. Students are asked to determine in what direction the last line of sight will lie and then to calculate the minimum tree radius that will make the center of the orchard a true hideout.

The subsequent discussion will introduce the phrase *last line of sight* and look at various methods the students may have used to calculate the tree radius. Finally, a rectangular coordinate system is established in which the center of the orchard is at (0, 0) and trees are planted at lattice points within the circle.

Approximate Time

5 minutes to introduce activity
25 minutes for activity (at home or in class)
10 to 15 minutes for whole-class discussion

Classroom Organization

Individuals, followed by whole-class discussion

Doing the Activity

Clarify the assignment by going over the diagram of the orchard "on a lot whose radius is 1 unit."

Discussing and Debriefing the Activity

Part I of this activity is dealt with in the presentations from *A Geometric Summary*. The discussion here is about Part II of the assignment.

Have some students restate or summarize the unit problem in their own words. Then have one or two students present their answers for each question of Part II.

Question 1
In Question 1, they should realize that the lines of sight that stay clear the longest are those at a 45° angle from the vertical and horizontal directions. You can refer to these directions as *last lines of sight*. (Students may describe these lines in terms of direction—for instance, "due northwest.")

Students should also see that these final lines of sight are not blocked until the trees "grow into one another."

Note: In the case of this mini-orchard of radius 1, once the trees have grown to block the lines of sight, Madie and Clyde cannot get out of their orchard at all, even by a nonstraight path. Until the activity *More Mini-Orchards,* students may not realize that, beginning with an orchard of radius 2, it is possible for all lines of sight to be blocked without "caging" Madie and Clyde inside the orchard.

Question 2
Question 2 requires students to find out how big the trees must get to block the last lines of sight. The key observation is that the trees must grow until they touch one another.

The diagram below illustrates that for two "adjacent" trees to touch each other, they must grow to the point at which each radius is half the distance between their centers. The diagram shows the distance between centers by the dashed line, labeled d, so the trees must grow until the radius is half of d.

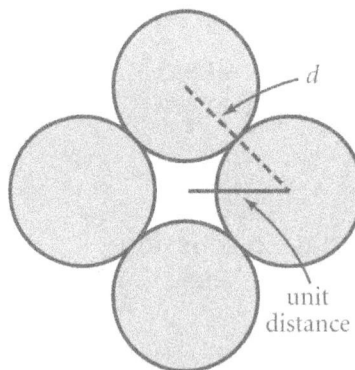

There are many methods by which students might find the length d (besides remembering it from previous work). Here are four.

- Applying the Pythagorean theorem: In the right triangle with hypotenuse d, $1^2 + 1^2 = d^2$, so $d = \sqrt{2}$.
- Using trigonometry: The right triangle with hypotenuse d has acute angles of 45°, so, for instance, $\sin 45° = \dfrac{1}{d}$ and $d = \dfrac{1}{\sin 45°}$.

- Using area: The square formed by connecting the centers of the four trees has an area of 2 square units (because it is made up of four half-squares), so its side (which is the desired diagonal) has length $\sqrt{2}$.
- Measuring: Students might get $d \approx 1.4$. This is a legitimate initial approach, but you should insist that they also find the exact value.

The Desired Radius

You may have to remind students that the answer to Question 2 is actually "half of *d*." That is, the orchard of radius 1 will become an orchard hideout when all of the trees grow to a radius of $\dfrac{\sqrt{2}}{2}$.

Introduce the term *hideout tree radius* to refer to the value $\dfrac{\sqrt{2}}{2}$. Emphasize that this value depends on the radius of the orchard itself. The value $\dfrac{\sqrt{2}}{2}$ applies only to the orchard of radius 1. Ask, Why is the answer not simply one-half of a unit?

Coordinatizing the Orchard

Ask, What would be a convenient way of referring to individual trees in the full-size orchard? Students may have various suggestions, but the goal here is to introduce a rectangular coordinate system for the orchard, with the center of the orchard at (0, 0), the east-west line through the center as the *x*-axis, and the north-south line through the center as the *y*-axis. Be sure to properly acknowledge any alternative ideas that students suggest, but if they do not come up with the idea of a rectangular coordinate system, you should introduce it yourself.

Be sure to include all four quadrants in your discussion, and have students identify points in each quadrant in terms of their coordinates.

Ask students to describe where the trees are located using this system—namely, at points with integer coordinates that are 50 units or less from (0, 0). (Clarify if needed that these points are the *centers* of the trees.) Introduce the term **lattice point** for a point in the coordinate system whose coordinates are integers.

Once the coordinate system has been described, you can suggest to students that graph paper may be useful in working with the unit problem.

Key Questions

Why is the answer not simply one-half of a unit?

What would be a convenient way of referring to individual trees in the full-size orchard?

Supplemental Activity

Right and Isosceles (reinforcement) asks students to find the exact values of the trigonometric functions for a 45° angle.

POW 1: Equally Wet

Intent

Students discover that the circumcenter of a polygon lies at the intersection of the perpendicular bisectors of the sides.

Mathematics

Students will be introduced to the terms **circumscribed** and **circumcenter**. They will learn that the intersection of the perpendicular bisectors of the sides of a polygon determines the circumcenter of the polygon, which is in turn the center of the circumscribed circle. Students will also practice translating a problem statement for a physical situation into mathematical language.

This POW is one of several assignments in *The Orchard Hideout* in which students are asked to describe the set of points satisfying a given geometric condition. In traditional high school geometry, these are called *locus* problems.

Progression

This POW asks students to determine how to place a sprinkler with a circular spray pattern so that two and then three flowers will get the same amount of water. It then prompts them to explore the same situation for four or more flowers.

Only Two Flowers will help students get started by leading them to discover that the perpendicular bisector is the locus of all points equidistant from the two end points of a line segment. *From Two Flowers to Three* helps them to explore a situation with three flowers placed at specific coordinates. The generalization of this situation is left for their POW write-up.

Following student presentations of the POW solution, discussion will confirm that students know how to locate the circumcenter of a polygon, and will introduce the terms **circumcenter** and **circumscribed**. Finally, a brief discussion will focus on how to phrase the problem situation in mathematical language.

Approximate Time

5 minutes to introduce the POW
1 to 3 hours for activity (at home)
35 minutes for presentations and discussion

Classroom Organization

Individuals, followed by whole-class discussion

Materials

Presenters will need presentation materials, such as transparencies and pens, a few days prior to the due date.

Doing the Activity

In *Only Two Flowers*, students will look at the first question of the POW—the two-flower case. In the subsequent discussion, they will prove that any point on the perpendicular bisector of a line segment is equidistant from the endpoints of the segment and discuss how this principle relates to the two-flower case. They will then need to consider the case of more than two flowers to complete work on this POW. Students may need to be reminded that they can't place the sprinkler and then adjust the flowers.

Point out that the "Problem Statement" portion of the write-up asks students to express the problem in mathematical terms. This idea will be discussed further following the activity *A Perpendicularity Proof*.

The day before presentations are scheduled, choose three students to make the POW presentations, and give them overhead transparencies and pens to take home to use for preparing the presentations. You may want to ask one of the presenters to begin with a review of results from *Only Two Flowers*.

Discussing and Debriefing the Activity

Ask the three students to make their POW presentations, perhaps beginning with a review of the two-flower case.

Three Flowers

For the general case, students will probably build on the result for two flowers, where they saw that the solution is to place the sprinkler on the perpendicular bisector. For three flowers, they might say that the solution is to place the sprinkler where the perpendicular bisectors all intersect.

You may have to bring out the idea that actually *three* perpendicular bisectors are involved. That is, if the flowers are at points P, Q, and R, then the sprinkler must be on the perpendicular bisector of $\overline{PQ}$, on the perpendicular bisector of $\overline{PR}$, and on the perpendicular bisector of $\overline{QR}$, as shown here.

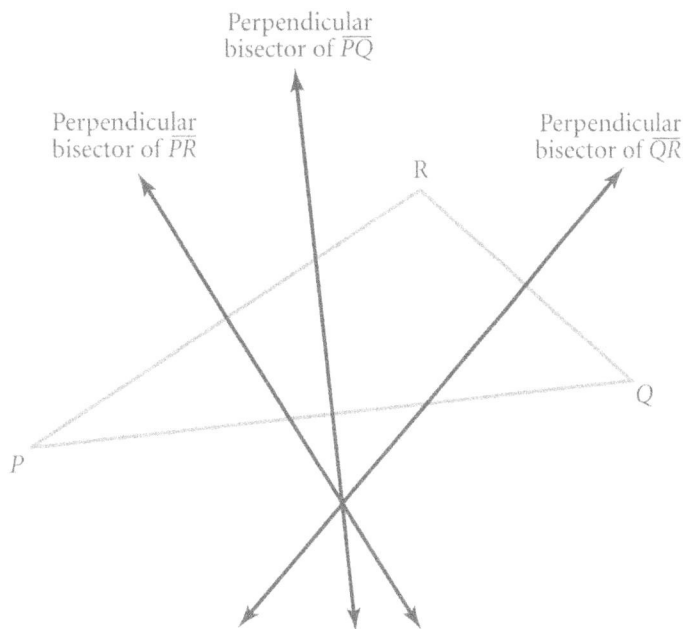

Students may not have considered the question of whether these perpendicular bisectors must meet. If it doesn't come up, you should raise the question. Using *P*, *Q*, and *R* as the locations of the flowers ask, Why should the perpendicular bisector of $\overline{PQ}$ meet the perpendicular bisector of $\overline{QR}$?

If you don't get a good response to this question, you may find it useful to broaden it, asking, Does every pair of lines intersect? This should elicit some mention of parallel lines, and you can then ask, Could the perpendicular bisectors be parallel?

Let students work on this question in groups until someone recognizes that if the original points are **collinear**, then the perpendicular bisectors will be parallel. Then, the sprinkler problem will have no solution.

Be sure students realize that for the case of three *noncollinear* points, there is a unique location for the sprinkler that is equidistant from all three, and that this point is the intersection of the three perpendicular bisectors.

You might ask students if they can summarize the result for the three-flower case using the phrase "if and only if." For instance, they might say, "There is a way to place the sprinkler equidistant from three given flowers if and only if the flowers are not collinear."

The Circumscribed Circle
Point out that three noncollinear points can be thought of as the vertices of a triangle, and ask, What is the relationship between the location of the sprinkler and the triangle formed by the flowers? As a hint, remind students that the flowers are all the same distance from the sprinkler, and ask what one calls the set of all points that are a given distance from a particular point.

These suggestions should help students determine that the sprinkler location is the center for a circle that goes through all three flowers. Tell students that this circle is said to be **circumscribed** about the triangle and that the center of this circle is called the *circumcenter* for the triangle.

POW 2: On Patrol involves the circle **inscribed** in a triangle. We recommend that you *not* mention this connection, but rather let students discover it on their own.

Four or More Flowers

For four or more flowers, it will only be possible to place the sprinkler the same distance from them all if they all lie on a common circle. If no one comes up with this generalization, you may want to let students work on the case of four flowers in groups, drawing on their discussion of the three-flower case. However, if there are no clear ideas about what to do in the case of more than three flowers, you can leave it unresolved for now.

Reviewing Problem Statements

When the discussion has been completed, you may want to ask for volunteers to read their problem statements so that you can help students understand how to phrase them in more mathematical terms. (See the subsection "The Problem Statement" in the discussion notes for *A Perpendicularity Proof*.). If students talked about flowers and sprinklers in POW 1, urge them to work on this issue on the next POW. After discussing the abstract mathematical formulation of the problem, you might ask students to suggest other contexts in which this issue might arise.

Next POW

POW 2: On Patrol is a follow-up to this POW. The activity *Down the Garden Path* builds on POW 1 and is also preparatory for POW 2, so it's important to have presentations for POW 1 before assigning the other two activities.

Key Questions

Why should the perpendicular bisector of $\overline{PQ}$ meet the perpendicular bisector of $\overline{QR}$?

Does every pair of lines intersect?

Could the perpendicular bisectors be parallel?

What is the relationship between the location of the sprinkler and the triangle formed by the flowers?

Supplemental Activities

Why Do They Always Meet? (extension or reinforcement) asks students to explain why the point of intersection of two of these perpendicular bisectors must also be on the third.

Inscribed Angles (extension) builds on the idea of circumscription by asking students to prove that an angle inscribed in a semicircle must be a right angle.

More Inscribed Angles (extension) builds further upon *Inscribed Angles*.

Angles In and Out (extension and reinforcement) uses the context of *Inscribed Angles* and *More Inscribed Angles* to introduce the statement that each exterior angle of a triangle is equal to the sum of the two nonadjacent interior angles, and then asks students to prove it.

Only Two Flowers

Intent

Students begin work on *POW 1: Equally Wet* by exploring the two-flower case in the first question.

Mathematics

Students will establish intuitively that the perpendicular bisector is the locus of all points equidistant from the endpoints of a line segment.

Progression

Students answer the first question from POW 1: where to place a sprinkler so that it will be exactly the same distance from each of two flowers. Subsequent discussion will identify all possible solutions as lying on the perpendicular bisector of the line segment connecting the two flowers. Students will be introduced to the terms **equidistant** and **perpendicular bisector**. In *A Perpendicularity Proof*, they will go on to prove this conjecture.

Approximate Time

20 minutes for activity (at home or in class)
10 minutes for discussion

Classroom Organization

Individuals, followed by whole-class discussion

Doing the Activity

Only Two Flowers refers to the first question of POW 1, so the POW needs to have been introduced before assigning this activity. The discussion following this activity needs to be completed before assigning *A Perpendicularity Proof*.

Discussing and Debriefing the Activity

You might begin the discussion by marking two points on the chalkboard to represent the flowers. We suggest that to emphasize the generality, you place the points so that the segment connecting them is neither vertical nor horizontal. You may find it useful to label the points *A* and *B* for convenience of discussion.

Where could the sprinkler be placed? How do you know this fits the condition? Ask several students to each come up and mark a point where the sprinkler could be placed. For the first two or three examples, you might specifically ask why the given point fits the condition stated in the problem, so that students focus on the idea that the point must be the same distance from the two initial points, *A* and *B.* Introduce the term **equidistant** in this context.

After a few examples, you might have a diagram like the one shown below. Continue until the pattern of the solutions becomes fairly clear. Ask, What do you see happening with these possible locations?, or wait until a student articulates that they form a straight line.

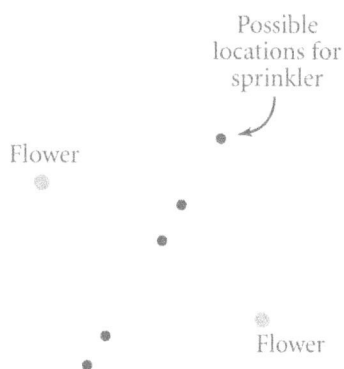

Descriptions of the Line: The Perpendicular Bisector

Now ask for verbal descriptions of this line. If students need some guidance, you might ask, How is this line related to $\overline{AB}$, the segment connecting the two flowers? Get students to identify two key elements.

- The line goes through the midpoint of $\overline{AB}$.
- The line is perpendicular to $\overline{AB}$.

Once these two characteristics have been identified, tell students that the line of solutions is called the **perpendicular bisector** of the line segment.

Bring out the idea that in identifying the perpendicular bisector as the solution to the homework, students are saying two things:

- Every point on the perpendicular bisector is equidistant from points *A* and *B.*
- Every point that is equidistant from points *A* and *B* is on the perpendicular bisector.

In other words, every point on the line is a solution, and every solution is on the line.

Key Questions

Where could the sprinkler be placed? How do you know this fits the condition?

What do you see happening with these possible locations?

How is this line related to $\overline{AB}$, the segment connecting the two flowers?

A Perpendicularity Proof

Intent

Students develop a proof for the conjecture they made in *Only Two Flowers.*

Mathematics

In *Only Two Flowers*, students discovered intuitively that every point on the perpendicular bisector of a line segment is equidistant from the endpoints of that segment. In this assignment they are challenged to prove that statement. Though students may use other ways to prove this, the discussion will include a review of the side-angle-side congruence principle. The concept of a **converse** is also introduced.

Progression

Each student writes a perpendicularity proof based on the diagram. In the discussion following this activity, various students can volunteer to share their proofs. In particular, be sure to look at both a proof that involves the Pythagorean theorem and a proof using the side-angle-side congruence axiom.

Students can then be asked to express the reverse of the statement from this activity in "If-then" format, introducing the concept of a *converse*. Strengthen their understanding of this concept by looking at the converse of the Pythagorean theorem.

Approximate Time

20 minutes for activity
20 minutes for discussion

Classroom Organization

Individuals, followed by whole-class discussion

Doing the Activity

Tell the class that the next task is to *prove* that every point on the perpendicular bisector is equidistant from A and B. Probably the simplest proof is one that uses the Pythagorean theorem, although students may find other approaches. You may need to help students "translate" from the fact that C is the midpoint of $\overline{AB}$ to the condition that the lengths CA and CB are equal, and from the fact that l is perpendicular to $\overline{AB}$ to the condition that triangles ACD and BCD are right triangles.

Discussing and Debriefing the Activity

Let volunteers share ideas. The Pythagorean theorem provides one excellent approach, and students should notice that triangles *ACD* and *BCD* are right triangles whose legs have the same pair of lengths. (In fact, the two triangles have a leg in common.) Therefore, by the Pythagorean theorem, the lengths of the hypotenuses of these two triangles must be equal.

Although the proof using the Pythagorean theorem is perhaps the easiest, there are other approaches, and you should elicit a variety of ideas. In particular, this is a good opportunity to discuss the following congruence principle, which is included in *A Geometric Summary*:

> **If two sides and the angle they form in one triangle are equal to the corresponding parts of another triangle, then the triangles must be congruent.**

This is often called the *side-angle-side congruence principle*.

If no one uses this approach in their discussion of the activity, ask, Are there any congruent triangles in the diagram? Once students identify the triangles *ACD* and *BCD* as congruent, ask how they know the triangles are congruent. How do you know they are congruent?

This is a subtle issue, because the congruence principle is often taken as an axiom. Rather than deal with the issue of axioms right away, begin with whatever intuitive explanations students have, perhaps focusing on the "rigidity" of triangles. That is, bring out that once the two lengths and included angle are specified, the triangle has no "flexibility."

After an intuitive discussion of this congruence principle, tell students that in formal work in geometry, this principle is taken as an *axiom*—that is, as an accepted truth about congruence.

The Converse

What is the hypothesis about *D* in *A Perpendicularity Proof*? What is the conclusion? (You may want to review this terminology and the structure of **"If-then" statements** for those who have not seen them before.)

Help the class to state these two conditions as concisely as possible. The hypothesis can be stated briefly as follows:

> **Point *D* is on the perpendicular bisector of the line segment connecting *A* and *B*.**

The conclusion is:

> **Point *D* is equidistant from *A* and *B*.**

Have the class put these two parts together in an "If-then" form. Thus, in the activity, students proved this statement:

> **If *D* is on the perpendicular bisector of the line segment connecting *A* and *B*, then it is equidistant from *A* and *B*.**

What does this mean in terms of Leslie's sprinkler dilemma? Help students articulate the idea that if Leslie places the sprinkler at any point on the perpendicular bisector, the flowers will be watered equally.

The discussion of that activity should lead to an additional result, namely, that Leslie can use *only* points on the perpendicular bisector. In other words, not only is every point on the perpendicular bisector a solution for Leslie, but every solution is on the perpendicular bisector.

Help the class to express this other result as an "If-then" statement. In terms of the diagram, it can be stated as:

> **If a point is equidistant from *A* and *B*, then it is on the perpendicular bisector of the line segment connecting *A* and *B*.**

Emphasize that this statement has the same two parts as the previous statement, but that the roles of **hypothesis** and **conclusion** have been reversed.

Tell the class that if the hypothesis and conclusion of one "if-then" statement are interchanged to form a second "if-then" statement, then each statement is called the **converse** of the other. In other words, each of the following two statements is the converse of the other.

- **If *D* is on the perpendicular bisector of the line segment connecting *A* and *B*, then it is equidistant from *A* and *B*.**
- **If *D* is equidistant from *A* and *B*, then it is on the perpendicular bisector of the line segment connecting *A* and *B*.**

Remind the students that in the activity *A Perpendicularity Proof,* they proved the first of these statements, but not the second. Tell them that proving the second statement is explored a supplemental activity, *The Perpendicular Bisector Converse*.

Some "Real-Life" Examples of the Converse

What are some real-life examples of "if-then" statements? Which are true? Which converses are true? If needed, give an example to get them started, such as "If an animal is a poodle, then it is a dog."

Have them identify the hypothesis and conclusion in each case, and then give the converse. For each example, discuss whether the original statement is true and whether its converse is true.

"If and Only If"

This is a good occasion to introduce the phrase **"if and only if."** You can simply tell students that mathematicians often use this phrase when having either of two conditions true guarantees that the other is also true. Review the two conditions and give students the following statement using "if and only if":

> ***D* is equidistant from *A* and *B* if and only if *D* is on the perpendicular bisector of the line segment connecting *A* and *B*.**

You can explain that this is mathematically equivalent to combining the two separate "if-then" statements.

The Pythagorean Converse

The Pythagorean theorem provides another familiar example of the concept of a converse. Ask, What does the Pythagorean theorem say? Students will probably say something like this:

If a triangle is a right triangle, then the lengths of its sides satisfy the equation $a^2 + b^2 = c^2$.

If students simply offer the equation, be sure to have them clarify the roles of the variables, making clear that a and b are the lengths of the legs and c is the length of the hypotenuse.

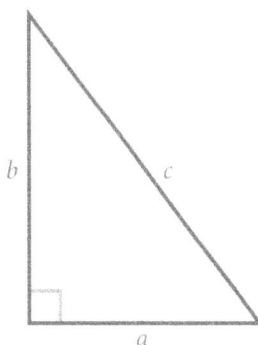

Ask, How can you use the Pythagorean theorem in reverse? They should see that the theorem can be stated in reverse:

If the lengths of the sides of a triangle satisfy the equation $a^2 + b^2 = c^2$, then the triangle must be a right triangle.

Identify this "reverse" direction as the converse of the Pythagorean theorem.

The Problem Statement

Ask, What would you give as the problem statement for the activity *Only Two Flowers?* Help students to extract the mathematical essence of the problem and to know that they need not include the flowers and the sprinkler. For example, the problem statement for the assignment might be, "Given two points *A* and *B*, what is the set of points that are equidistant from *A* and *B*?"

Describing the Solution

Students might have answered the question in *Only Two Flowers* by saying, "Place the sprinkler anywhere that is equidistant from the two flowers." Tell them that this is not an adequate answer because it doesn't explain how to find these locations.

They can now answer that problem more precisely by saying, "Place the sprinkler anywhere on the perpendicular bisector of the segment connecting the two flowers." Tell them that they should strive for similar precision in answering the more general question raised in the POW.

Key Questions

Are there any congruent triangles in the diagram? How do you know they are congruent?

What is the hypothesis about *D* in *A Perpendicularity Proof*? What is the conclusion?

What does this mean in terms of Leslie's sprinkler dilemma?

What are some real-life examples of "if-then" statements? Which are true? Which converses are true?

What does the Pythagorean theorem say?

How can you use the Pythagorean theorem in reverse?

What would you give as the problem statement for the activity *Only Two Flowers?*

Supplemental Activities

The Perpendicular Bisector Converse (extension) asks students to prove the converse of the statement that they proved in *A Perpendicularity Proof.*

From Two Flowers to Three

Intent

Students continue work on *POW 1: Equally Wet* with a specific example of the case of three flowers.

Mathematics

In this assignment, students build on their discovery in *Only Two Flowers* to further develop the foundation they will need to ultimately conclude in their POW that the circumcenter of a polygon lies at the intersection of the perpendicular bisectors of the sides, if such a point exists. In this case, they locate a point equidistant from the vertices of a given triangle by finding the intersection of the perpendicular bisectors of two sides of the triangle.

Progression

Students are given the coordinates of three non-colinear points and are asked to find a point equidistant from all three points. In the discussion afterwards, emphasize that there is a unique solution, but leave the generalization as part of the POW.

Approximate Time

25 minutes for activity (at home or in class)
10 minutes for discussion

Classroom Organization

Individuals, followed by whole-class discussion

Discussing and Debriefing the Activity

Let a volunteer present the solution to the problem. We recommend that you keep the discussion focused on the specific example, because students should develop the generalization for their POW (along with consideration of the cases of more than three flowers).

In the specific example, two of the sides of the triangle formed by the flowers are horizontal and vertical. Because of this, students will probably be able to get coordinate or equation descriptions of two of the perpendicular bisectors and see that they meet at (9, 5). Emphasize that this is the *only* place where Leslie can place the sprinkler in this case. Raise—*but do not resolve*—the question of whether there is always a unique solution in the case of three flowers.

Note: Some students may notice that the solution, (9, 5), is the midpoint of the hypotenuse of the right triangle. During the unit, students will develop the tools to prove the generalization of this observation, which is discussed in the supplemental activities *Right in the Center* and *Hypotenuse Median*. However, students are not ready for either of these supplemental assignments at this point.

More Mini-Orchards

Intent

Students discover that an orchard can become a hideout even before the tree trunks reach one another.

Mathematics

In this activity, students study the cases of orchards of radius 2 and 3. Students are not expected to be able to calculate exact values for the hideout tree radius for orchards of radius 2 and 3, as they are challenged to do in Question 3. Rather, the goal of this activity is simply for students to figure out from their sketches that it is not necessary for the trees to be touching in order to block the line of sight. Discussion of this principle will introduce the idea of a line being **tangent** to a circle.

Progression

Students are asked in the first two questions to sketch orchards of radius 2 and 3, and to approximate the hideout tree radius for each orchard. Question 3 challenges more enterprising groups to find exact values for these radii. The subsequent discussion focuses on ensuring that students understand that the line of sight becomes blocked before the trees grow to such a size that they touch one another. Examining a line of sight that is neither horizontal nor vertical makes a good opportunity to introduce the concept of tangency, which will be important in the eventual solution of the unit problem.

Approximate Time

20 minutes for activity
15 minutes for discussion

Classroom Organization

Individuals or small groups, followed by whole-class discussion

Materials

Large-grid graph paper or *1-Inch Graph Paper* blackline master
More Mini-Orchards: Radius 2 blackline master
More Mini-Orchards: Radius 3 blackline master *More Mini-Orchards: Gaps for Radius 3* blackline master

Doing the Activity

Have students sketch mini orchards, first with radius 2 and then with radius 3. Students can represent the trees as small dots. You might suggest that they draw a sequence of diagrams, first with very small dots and then with larger dots, gradually increasing the radius of the dots until the center appears to be a hideout.

The main goal is for students to realize that the tree trunks do not have to touch one another to create a hideout. (This may be easier for them to see using the orchard of radius 3 than the orchard of radius 2.)

You may find it helpful to interrupt group work on the activity after some exploration and use transparencies of the radius 2 and radius 3 blackline masters. (These masters do not show lines of sight.)

Expectations for Question 3

Question 3 is primarily intended as a challenge for groups that finish Questions 1 and 2, and you should not expect students at this stage to get the exact answers. More likely, they will be able to determine that in the orchard of radius 2, a tree trunk radius smaller than $\frac{1}{2}$ suffices to create a hideout, but they may be unable to find a particular smaller value that works. They may reach a similar conclusion for the orchard of radius 3.

Discussing and Debriefing the Activity

Let one or two students present their groups' ideas about Question 1. They should certainly all agree that there are 12 trees (two in each direction on each axis, and one in the interior of each quadrant). Be sure to use a diagram like the one below, showing the location of the 12 trees.

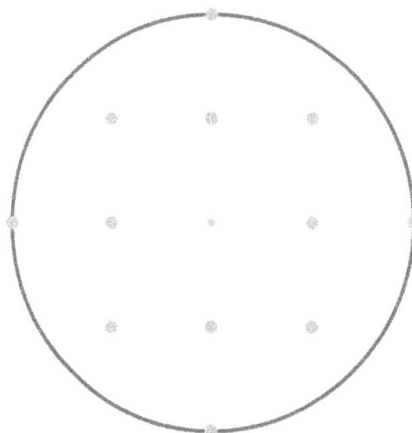

Orchard of radius 2 shortly after planting
(Also in the blackline master for circles of radius 2)

In determining what tree trunk radius is needed to make this mini-orchard into a hideout, students may have varying levels of understanding. Let one or two students present their ideas, and have others offer additional thoughts.

You can use a sequence of diagrams like those below (together with the initial diagram) to show the gradual growth of the trees. Students may find it illuminating to see the transparencies superimposed one on another as the trees grow.

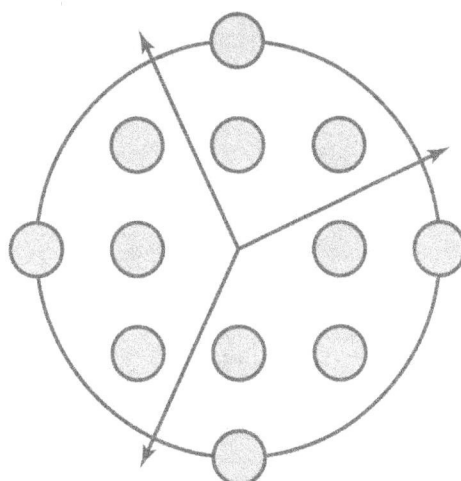

Tree trunks have radius $\frac{1}{4}$, and there are many lines of sight from the center out of the orchard.

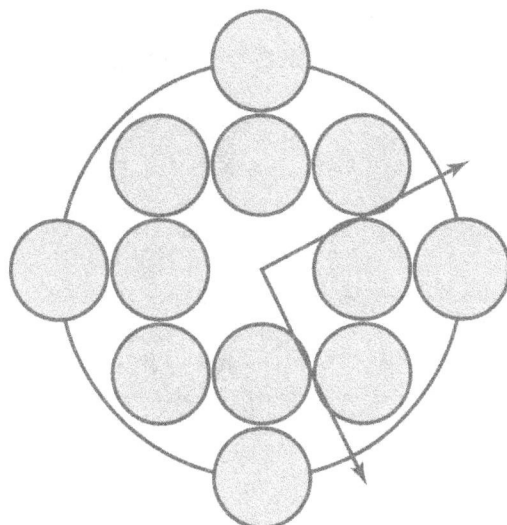

Even with tree trunks of radius slightly less than $\frac{1}{2}$, the center is a true hideout. Every line of sight is blocked by some tree.

The second diagram of this sequence shows that a tree trunk radius of $\frac{1}{4}$ is not enough to make this orchard a hideout, so the hideout tree radius for an orchard of radius 2 is somewhere between $\frac{1}{4}$ and $\frac{1}{2}$.

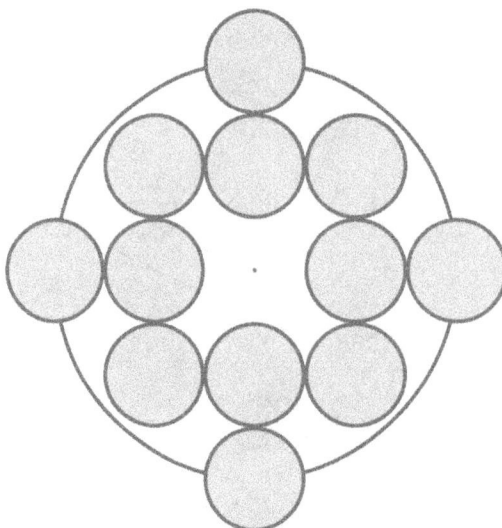

When the tree trunks reach a radius of $\frac{1}{2}$,
the center of the orchard is "caged."

Some students may not yet be convinced that a radius of less than 2 will suffice to make the orchard into a hideout. But they should recognize that for an orchard on a lot of radius 2 (or more), if the radius of the tree trunks is equal to $\frac{1}{2}$, then the trees will be growing into one another. In other words, as soon as the trees reach a radius of $\frac{1}{2}$, it will no longer be possible to see out of the orchard from the center. In fact, it will no longer be possible to get out of the orchard at all.

Some groups may realize that a smaller tree trunk radius will suffice for the lot of radius 2. The diagrams indicate this, but it may not be completely clear to students. If they don't come up with this idea, you may want to wait and see what they do with the case of an orchard of radius 3 and then return to the case of radius 2 (as discussed below under *Lines of Sight*).

The Orchard of Radius 3
A similar discussion should take place for the mini-orchard of radius 3. In this case, however, it should be fairly clear to students that the trees do not have to grow into one another for the orchard to become a hideout.

You can use a sequence of diagrams like those below. (Like the diagrams for radius 2, these are included in larger versions, without the sight lines, in the *More Mini-Orchards: Radius 3* blackline master.)

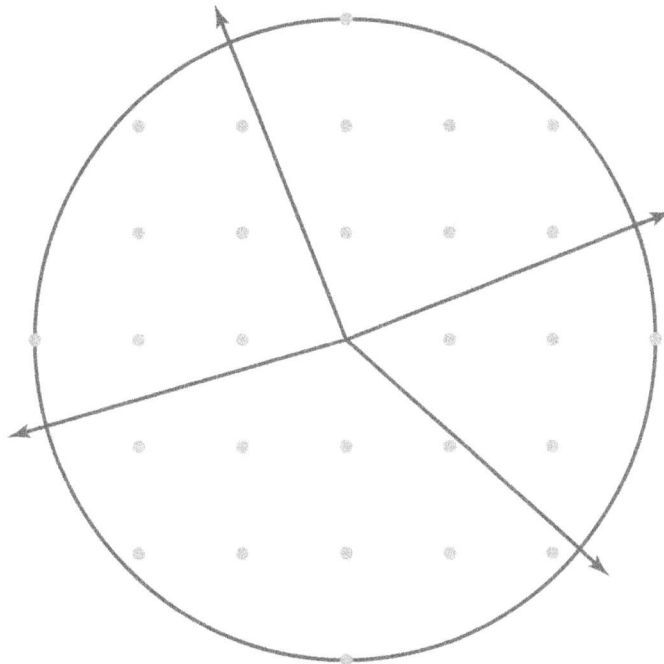

With newly planted trees, the orchard of radius 3 has many lines of sight out.

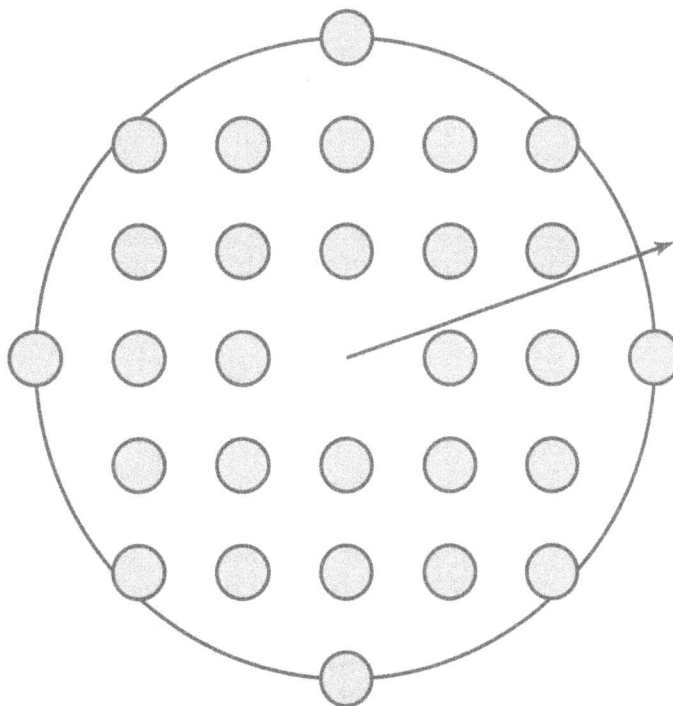

The trees are larger, with trunk radius $\frac{1}{4}$, but there are still lines of sight out.

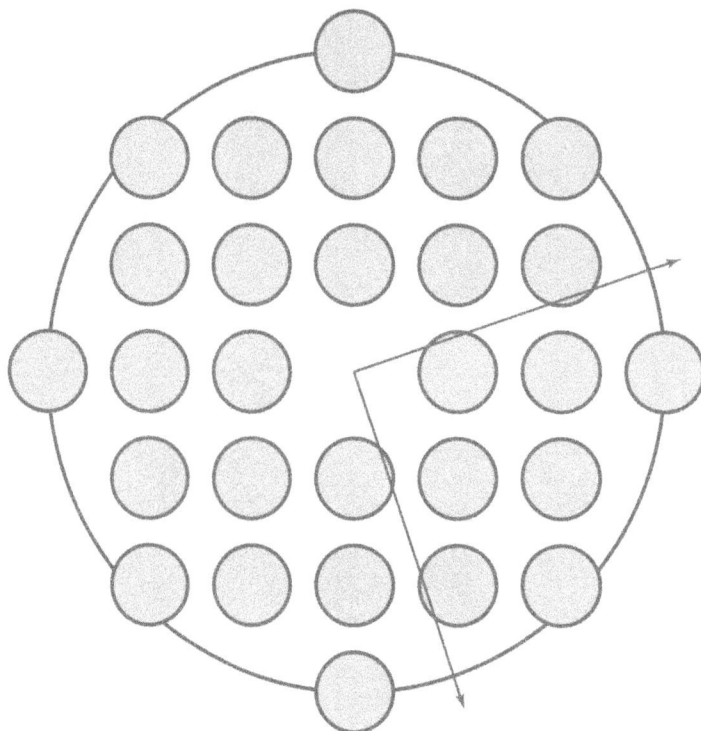

*The trees don't reach each other, but
there are no longer any lines of sight out.*

Summary

By the end of this discussion, students should realize that there is no easy answer to the orchard problem. They should be wondering what the exact smallest radius of the tree trunks is that will suffice to hide the center of the orchard. Tell them that answering this question is the main goal of the unit and that they will need to learn more about geometry to achieve this goal.

Lines of Sight

Note that the diagrams depict the "lines of sight" as rays emanating from the origin, since these rays fit the context better than lines do. For simplicity of language, however, we refer to them simply as lines. The geometry is the same in either case.

In *Geometry and a Mini-Orchard,* students saw that for the orchard of radius 1, Madie and Clyde would be able to see out for the longest amount of time if they looked in a direction at a 45° angle to the axes, such as due northeast.

Pose a similar question for the orchards of radius 2 and 3 by asking, Where should Madie and Clyde look in order to be able to see out for the longest time? With the case of radius 2, bring out that because of symmetry, all the "gaps" are essentially the same, although students may not know exactly which angle is best.

Then move on to the case of radius 3. Help students to realize that there are two "families" of gaps, which are illustrated by the two shaded areas in the next diagram.

- Gaps such as that between (1, 0) and (2, 1) (the lightly shaded area)
- Gaps such as that between (2, 1) and (1, 1) (the darkly shaded area)

One line of sight is shown within each gap.

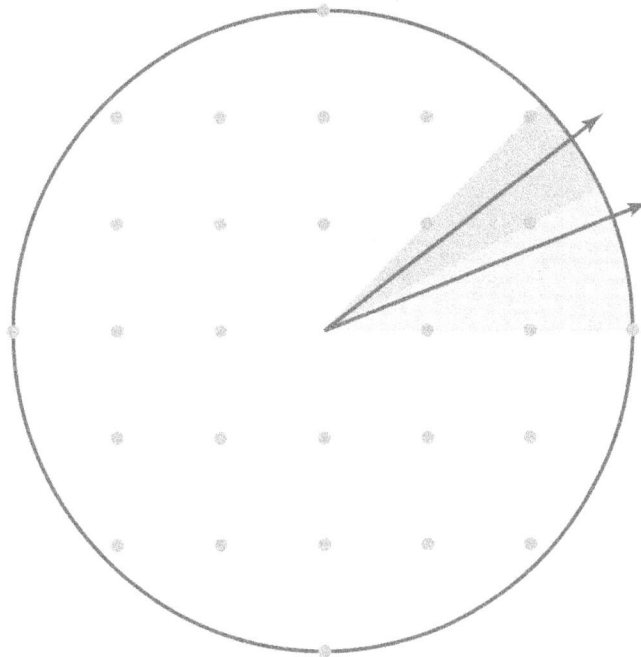

Again, students will not yet be able to determine exactly what angle is best for the line of sight, but intuition likely will suggest that a gap like that between (1, 0) and (2, 1) (the lightly shaded area) is larger than the gap between (2, 1) and (1, 1) (the darkly shaded area) and so will remain open longer.

Back to Radius 2

If students did not determine earlier that a tree trunk radius of less than $\frac{1}{2}$ would make the orchard of radius 2 a hideout, you can go back to that now. They may still not be sure what the hideout tree radius is, but they should be able to figure that something less than $\frac{1}{2}$ will work.

You can illustrate this using the diagram of the radius 2 mini-orchard with tree trunks of radius $\frac{1}{2}$. Draw the (blocked) line of sight from the origin through $(1, \frac{1}{2})$, as shown here, and ask where this line hits the trees centered at (1, 0) and (1, 1).

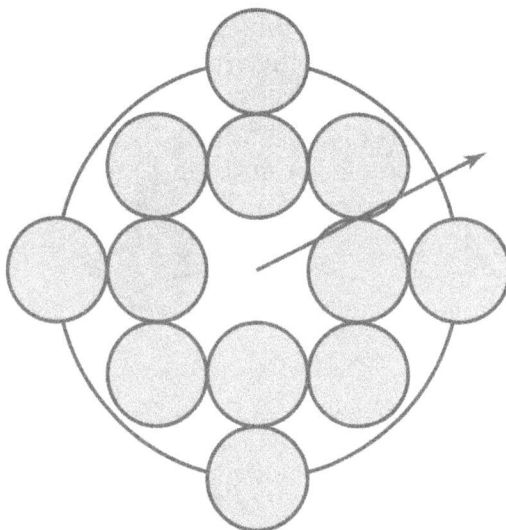

The crucial idea to bring out is that this line not only touches both trees at $(1, \frac{1}{2})$

but intersects each tree at other points as well. Therefore, the trees could actually be slightly smaller than this and still block this line of sight.

Tangent to a Circle

The preceding discussion of the orchard of radius 2 provides a good context for introducing the idea of a **tangent** to a circle. (If that discussion took place on a previous day, you may want to review it now as an introduction to the concept of tangents.)

Ask students, *What do you call a line that touches a circle at only a single point?* You can use one of the orchard diagrams to illustrate the idea.

If they don't know, introduce the term **tangent to a circle.** Point out that this is a different (but related) use of the word *tangent* from its trigonometry meaning.

Bring out that the line of sight from the origin through $(1, \frac{1}{2})$ is *not* tangent to the

circle of radius $\frac{1}{2}$ with center at $(1, 0)$.

There is a connection between the use of the word *tangent* in the sense of *tangent to a circle* and the use of the word *tangent* in trigonometry. In the accompanying diagram, if the circle has radius 1, then $\overline{AB}$, which is part of the line tangent to the circle at A, has length equal to $tan\theta$.

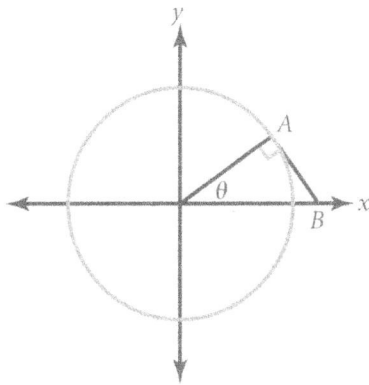

Later in the unit (in *Perpendicular and Vertical*), students will prove that a tangent to a circle and the radius to the point of tangency are perpendicular.

Key Questions

Where should Madie and Clyde look in order to be able to see out for the longest time?

What do you call a line that touches a circle at only a single point?

In, On, or Out?

Intent

Students explore the relationship between the Pythagorean theorem and the definition of a circle.

Mathematics

The students' use of the Pythagorean theorem in this activity is used as a foundation in the subsequent discussion to develop the equation of a circle centered at the origin. This discussion also makes an ideal opportunity to review the use of "if-then" and "if and only if" statements.

Progression

The students are given the coordinates of a number of points and are asked to determine whether each point lies inside, on, or outside a radius 10 circle centered at the origin. Although the students may use a number of approaches to obtain answers for most of the points, the discussion will focus on those points that appear to be on or very nearly on the circle. It is necessary to use the Pythagorean theorem to determine exactly where those points lie.

Once the students understand how the Pythagorean theorem is utilized to answer the activity's questions, this process will be generalized in order to develop the equation of a circle. The students can then be asked to restate the process required for *In, On, or Out?* in terms of this general equation and its related inequalities.

Approximate Time

5 to 10 minutes for introduction
30 minutes for activity (at home or in class)
50 minutes for discussion

Classroom Organization

Individuals, followed by whole-class discussion

Materials

Graph paper

Doing the Activity

If assigning this as homework, you may want to sketch an orchard of radius 10 in class.

Discussing and Debriefing the Activity

Let students discuss the examples in their groups. They should try to reach consensus about the various points under consideration. Ask, What do the trees on the boundary of the orchard have in common?

After the groups have had some time to share results, let students from various groups report on the group's decision about various sets of points. [You might have one group report on the points (11, 0), (10, 0), and (10, 1); another report on (9, 3), (9, 4), (9, 5), and (9, 6); a third report on (8, 5), (8, 6), and (8, 7); and a fourth report on (7, 6), (7, 7), and (7, 8).]

Some students may have completed the assignment by drawing diagrams of the orchard on graph paper, perhaps using a compass to draw the lot. If so, they needed to examine the lattice points, one by one, to decide which are and which are not within the circular lot. They will perhaps be uncertain about some of the points, depending on the quality of their compass and the sharpness of their pencil point.

Other students may have applied the Pythagorean theorem to find the distances of various points from the origin. Even so, they may not have a clear enough concept of the definition of a circle to fully understand this approach.

If no group is able to give a clear explanation based on the Pythagorean theorem, ask, How can you be sure about (7, 7) (which is barely inside the circle) or (8, 6) (which is on the circle), when your diagrams are somewhat imprecise?

Then ask, What exactly is a circle of radius 10? Get them to express this as clearly as they can, and post a formal definition like this one:
A circle is the set of all those points that are a given distance (the *radius*) from a given point (the *center*).

Students often are not clear on the formal distinction between a *circle* and its *interior* (its "inside"). Although the word *circle* is sometimes used in everyday speech to include the inside, students should know that the technical definition of a circle is only the *boundary* portion.

If needed, have students work on this some more in their groups after clarifying the definition of *circle*. Ask, How can you use right triangles and the Pythagorean theorem to determine where points are in relation to the circle? Have them illustrate this with a diagram making the right triangle and the use of the Pythagorean theorem explicit.

For example, for the point (8, 6), they might have a diagram like this.

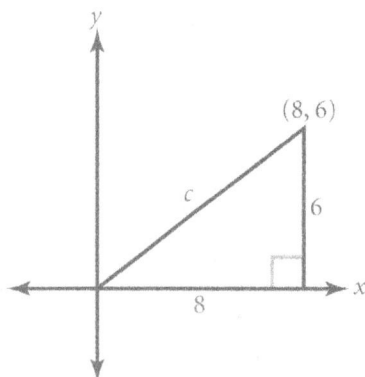

They might explain that the sides of the right triangle satisfy the equation $8^2 + 6^2 = c^2$, so $c = 10$, which means that (8, 6) is exactly on the boundary of the orchard.

Ask students for another term for the boundary of a circle. If necessary, tell them that it is called the **circumference.** Clarify that this term is used both for the *length* of the boundary and for the boundary itself. (The word *perimeter* has a similar dual use for polygons, and the word *radius* refers to both a line segment from the center to the circumference and the length of such a segment.) Bring out that circumference really is simply a special kind of perimeter.

If you have had difficulty getting students to use the Pythagorean theorem, you should probably review it with specific examples, including points *inside* and points *outside* the orchard.

The Equation of the Circle
Once students seem comfortable working with specific examples, ask them what equation they can write, using the coordinate variables x and y, so that the circle itself (that is, the *circumference* or *boundary*) will be the graph of that equation. What equation has this circle as its graph?

As needed, go back to the examples to bring out that the appropriate equation is $x^2 + y^2 = 10^2$ (or an equivalent, such as $\sqrt{x^2 + y^2} = 10$).

Emphasize that, as with all equations and their graphs, saying that the circle is the graph of this equation means two things.
If the coordinates of a point fit the equation, then the point is on the circle.
If a point is on the circle, then its coordinates fit the equation.
(This is another opportunity to use the term *converse*.)

Bring out that the equation $x^2 + y^2 = 10^2$ applies to *all* points on the circle, not only lattice points. You may want to have students come up with some nonlattice points that fit the equation. (It might be helpful to give them the value of x and ask for a value of y that would work.)

The Inside and Outside of the Circle

Ask, How can you use algebra to describe the set of points *inside* the circle? They should realize that this set is the graph of the *inequality* $x^2 + y^2 < 10^2$. Similarly, help them to articulate that the points *outside* the circle are precisely those that satisfy the condition $x^2 + y^2 > 10^2$.

Analogy with Straight Lines and Half Planes

Ask students what else they know about equations and related inequalities. As needed, remind them of their previous experiences with linear equations and inequalities. Bring out that the graph of a linear equation is a straight line and that the graphs of the two corresponding inequalities (replacing the equal sign with either of the inequality symbols) are half planes.

You might bring out that a circle, like a line, divides the plane into two portions.

Generalizing the Radius

Next, ask students to generalize from the case of radius 10 to the case of radius r. They should come up with something equivalent to this.
- A point (x, y) is on the circle if $x^2 + y^2 = r^2$.
- A point (x, y) is inside the circle if $x^2 + y^2 < r^2$.
- A point (x, y) is outside the circle if $x^2 + y^2 > r^2$.

Reminder: We are dealing here with circles with center at (0, 0).

More Work With "If-Then" and "If and Only If"

This is another good opportunity for students to apply the phrases "if-then" and "if and only if." Students should gradually become familiar with these elements of precise mathematical language. The best way to do this is for you to use the terminology and encourage them to do so as well. You need to clarify what the phrases mean and don't mean as you use them, but it is not fruitful to *teach* these phrases out of context.

For instance, you can use statements such as these in relation to the circle with center (0, 0) and radius r.
- If the coordinates of a point satisfy the condition $x^2 + y^2 = r^2$, then the point is on the circumference of the circle.
- If the coordinates of a point do not satisfy the condition $x^2 + y^2 = r^2$, then the point is not on the circumference of the circle.
- A point is on the circumference of the circle if and only if its coordinates satisfy the condition $x^2 + y^2 = r^2$.

You may want to bring out that the second of these statements is equivalent to the "only if" part of the third statement.

Key Questions

What do the trees on the boundary of the orchard have in common?

How can you be sure about (7, 7) (which is barely inside the circle) or (8, 6) (which is on the circle), when your diagrams are somewhat imprecise?

What exactly is a circle of radius 10?

How can you use right triangles and the Pythagorean theorem to determine where points are in relation to the circle?

What equation has this circle as its graph?

How can you use algebra to describe the set of points *inside* the circle?

Supplemental Activities

Not Quite a Circle (extension) builds on the students' work with the equation of a circle.

What's a Parabola?, Creating Parabolas, Coordinate Ellipses and Hyperbolas, Another View of Ellipses and Hyperbolas, Ellipses and Hyperbolas by Points and Algebra, Generalizing the Ellipse, and *Moving the Ellipse* (extensions) extend students' work in coordinate geometry to conic sections.

Coordinates and Distance

Intent

The activities in *Coordinates and Distance* provide a structure within which students develop the distance and midpoint formulas.

Mathematics

In *In, On, or Out?* students used the Pythagorean theorem to determine where trees lay with respect to the circle of the orchard boundary. This application of the Pythagorean theorem continues, leading to a natural development of the distance formula. In this manner, the distance formula and its roots in the Pythagorean theorem become firmly linked in students' minds, so that the distance formula is viewed as a simple application of the Pythagorean theorem rather than yet another formula that needs to be memorized.

The theme of distances within the orchard (coordinate plane) continues as students develop the midpoint theorem and verify that the midpoints they identify are indeed equidistant from the given points.

Development of a general plan for solving the unit problem establishes a need to understand the relationship between the radius, area, and circumference of a circle, motivating the activities that follow later in *All About Circles*.

Progression

Other Trees returns students to consideration of trees that lie on the boundary of a circular orchard, similar to what they just worked with in *In, On, or Out?* The discussion of that activity moves from the geometric symmetry of the coordinates of trees lying on the circle to how that symmetry is reflected in the algebraic representation of points on the circle. This prepares students to understand why it is not important which point is chosen as (x_1, y_1) in the distance formula. In *Sprinkler in the Orchard* and *The Distance Formula*, students continue to use the Pythagorean theorem to find distances between points on the coordinate plane and then generalize their work to develop the distance formula. This development of coordinate geometry is furthered through discovery of the midpoint theorem in *A Snack in the Middle*.

How Does Your Garden Grow? asks students to develop a general plan for solving the unit problem.

Other Trees
Sprinkler in the Orchard
The Distance Formula
How Does Your Garden Grow?
A Snack in the Middle

Other Trees

Intent

Students explore the symmetry of the coordinates of points that lie on a circle in the four quadrants.

Mathematics

This activity brings out the idea that each boundary tree leads to other trees using symmetry. The symmetry in the diagram is then related to the symmetry in the equation for a circle.

Progression

Students are challenged to find the coordinates of all of the boundary trees in an orchard where a tree at (5, 12) lies on the boundary, and then to generalize this for an orchard of any size. The ensuing discussion ensures that students not only notice the symmetry inherent in the diagram of the orchard, but that they also see why this symmetry is dictated by the equation of the circle.

Approximate Time

25 minutes for activity (at home or in class)
10 minutes for discussion

Classroom Organization

Individuals or small groups, followed by whole-class discussion

Discussing and Debriefing the Activity

You may wish to tell students that for Question 1, there are 12 points on the boundary, then challenge them to pool ideas in their groups to try to match this result. One member of a group can report on the group's findings to the entire class.

Students might use symmetry to see that having (5, 12) on the boundary means that the points (12, 5), (–5, 12), (–12, 5), (–5, –12), (–12, –5), (5, –12), and (12, –5) are also on the boundary. They will probably have to use the Pythagorean theorem to see that the points (13, 0), (0, 13), (–13, 0), and (0, –13) are on the boundary as well. Use the word *symmetry* in connection with the first group of points if students do not use it themselves.

Question 2

In trying to generalize the problem, students will probably see that if (a, b) is on the boundary of an orchard, then the points (b, a), $(-a, b)$, $(-b, a)$, $(-a, -b)$, $(-b, -a)$, $(a, -b)$, and $(b, -a)$ must also be on the boundary, as illustrated here.

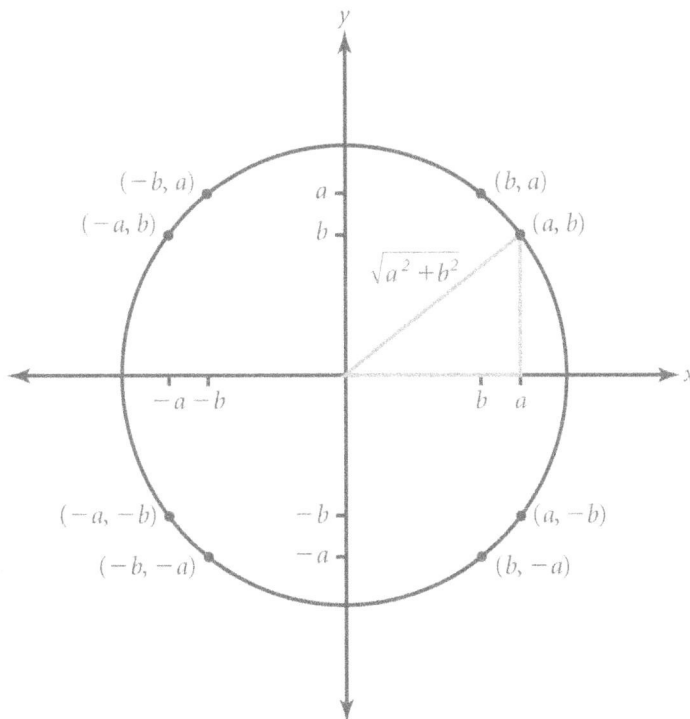

If they stop there, ask, *Must there be points similar to (13, 0)?* As needed, ask *Where did 13 come from in Question 1?*

Some students may suggest that points such as $(\sqrt{a^2 + b^2}, 0)$ need not be lattice points. If this comes up, acknowledge that this is a valid observation, ask for a pair of numbers for which $\sqrt{a^2 + b^2}$ is not an integer, and let the class ponder this.

You can then bring out the understanding that the expression $\sqrt{a^2 + b^2}$ does represent the radius of the orchard, because (a, b) is on the boundary. If the radius is an integer, then $\sqrt{a^2 + b^2}$ is also an integer.

Symmetry and the Equation

Ask students how they can use the equation of the circle (as developed in the discussion of *In, On, or Out?*) to show that these other points must be on the boundary. *What is it about the equation that tells us that if (a, b) is on the circle, then so is $(-a, -b)$?* You might begin by getting them to recall that the equation is $x^2 + y^2 = r^2$, where r is the radius of the circle.

Have students express their explanations explicitly in terms of the equation. For instance, they might point out that if (a, b) is on the boundary, then it is a solution to the equation, which means $a^2 + b^2 = r^2$. Because $b^2 + a^2 = a^2 + b^2$, (b, a) is also a solution to the equation, and so (b, a) is also on the circle.

You need not go through this argument for all cases, but be sure to include a case involving change of sign in order to illustrate that $(-a)^2$ has the same numerical value as a^2. You might also want to show that substituting $x = \sqrt{a^2 + b^2}$ and $y = 0$ fits the equation $x^2 + y^2 = r^2$ as well.

Key Questions

Must there be points similar to (13, 0)?

Where did 13 come from in Question 1?

What is it about the equation that tells us that if (a, b) is on the circle, then so is $(-a, -b)$?

Supplemental Activity

Counting Trees (extension) leads students to discover some interesting facts about the number of boundary trees in an orchard with an integer radius.

Sprinkler in the Orchard

Intent

Students use the Pythagorean theorem to find the distance between pairs of points on the coordinate plane.

Mathematics

Using the Pythagorean theorem to find distances will prepare students to develop the distance formula in *The Distance Formula*. Laying this groundwork thoroughly will contribute to students recognizing the distance formula as a natural extension of the Pythagorean theorem, rather than as an abstract formula to be once memorized and then vaguely recalled. Because these distances are once more presented as the radii of circles, this also continues to build the foundation for the introduction of the more generalized equation of a circle, no longer centered at the origin, in *Defining Circles*.

Progression

Question 1 provides more practice for finding a distance from the origin using the Pythagorean theorem. Question 2 requires students to build on that skill to find distances when neither of the points are located at the origin. Question 3 is a challenging problem for groups that move very quickly through the first two questions. Depending on how well the students make the transition from the situation in Question 1 to that in Question 2, discussion may be unnecessary.

Approximate Time

30 minutes for activity
5 to 10 minutes for discussion

Classroom Organization

Small groups or individuals, followed by whole-class discussion (if needed)

Doing the Activity

Questions 1 and 2 of this activity primarily reinforce previous ideas and also serve as a lead-in to the distance formula, which students will develop in *The Distance Formula*.

Question 3, on the other hand, is quite difficult and is intended only for groups that move quickly through the first two questions.

Discussing and Debriefing the Activity

If time allows, you can have a few students present their work on Questions 1 and 2 or you can incorporate these examples into the discussion of *The Distance Formula*. You need not have any discussion of Question 3 and might even treat the problem as an ongoing challenge.

Once students realize that they need to subtract coordinates to find the sides of the right triangles involved, the activity becomes an exercise in using the Pythagorean theorem. You may find that no presentations are necessary because students have dealt with these issues in their groups.

Keep in mind that students will work on more problems like these for *The Distance Formula* and that the generalization will be discussed in detail following that activity.

The Distance Formula

Intent

Students develop the distance formula.

Mathematics

The distance formula determines the distance between any two points on the coordinate plane. The formula has underpinnings in the Pythagorean theorem, so this activity builds on student work in *In, On, or Out?* and *Sprinkler in the Orchard*.

Progression

This activity asks students to again find the distance between specific pairs of points on the coordinate plane, this time reflecting on and describing in detail how they did so. They are then asked to generalize this to create a formula for the distance between (x_1, y_1) and (x_2, y_2). The subsequent discussion should proceed at a suitable pace to allow students to make this transition on their own, from finding a specific distance to developing a general formula, if they have not done so already.

Approximate Time

25 minutes for activity (at home or in class)
10 to 15 minutes for discussion

Classroom Organization

Individuals, followed by whole-class discussion

Discussing and Debriefing the Activity

You can begin the discussion by having one or two students share their answers for Questions 1b and 2b, which are their verbal descriptions of how they found the distances asked for in Questions 1a and 2a.

Then have a student present Question 3. By building on the examples, students should be able to develop and explain the general **distance formula**

$$d = \sqrt{\left(x_2 - x_1\right)^2 + \left(y_2 - y_1\right)^2}$$

or an equivalent, such as the equation

$$d = (x_2 - x_1)^2 + (y_2 - y_1)^2$$

or give a clear set of verbal instructions, which you can then help them turn into such a formula. Be sure they explain their work with diagrams such as this one.

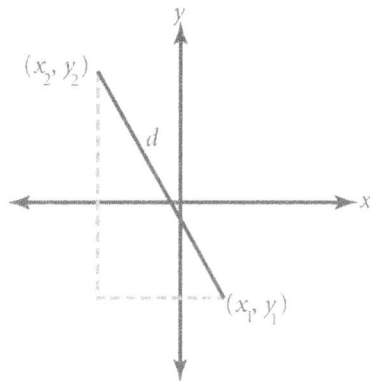

Note: It is not crucial that students memorize this formula. What is essential is that they understand the underlying role played by the Pythagorean theorem and right triangles in finding the distance between two points.

Hints on Developing the General Formula

If students are having trouble developing the general formula, you might have them find a formula for the distance from a general point (x, y) to a specific point, say, (3, 2). If that doesn't help, have them find the distance from various specific points to (3, 2) and continue to verbalize the process.

Diagrams showing the relevant right triangles should be part of these explanations. For instance, to find the distance from (1, 6) to (3, 2), students should provide a diagram like this one.

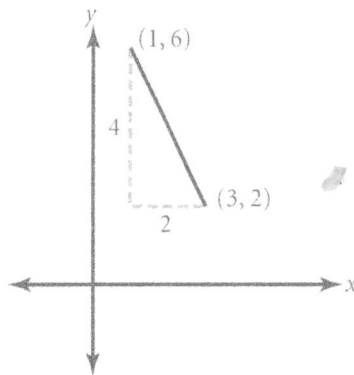

As needed, focus first on the process of finding coordinate differences to get the lengths of the legs of the right triangle and then on how to use those lengths to find the hypotenuse. Continue until someone can give the formula

$$d = \sqrt{(x-3)^2 + (y-2)^2}$$

or something equivalent, such as the equation

$$d^2 = (x-3)^2 + (y-2)^2$$

as well as an explanation using a diagram like the one below. (You might briefly review the role of *absolute value* here.)

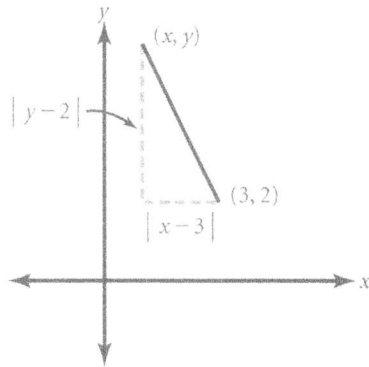

You might ask students to write a formula, as well as give an explanation, for the distance from a general point to another specific point. This may be a good time to introduce Question 4, by using something like (−1, −2) as the new specific point, to see if students know how to handle negative coordinates. A diagram might look like this.

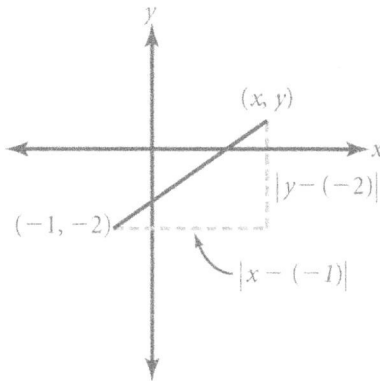

You can then return to the original Question 3.

Question 4

Question 4 requires clarification of two issues. First, whatever the signs of the individual coordinates, the horizontal leg of the related right triangle has a length of either $x_1 - x_2$ or $x_2 - x_1$, and similarly for the vertical leg. You may want to have students examine various sign combinations to see this. For example, bring out that the distance from (5, 3) to (−7, 3) can be found by computing the difference 5 − (−7).

This is a good opportunity to review the concept of absolute value. Ask, How can you express the length of the horizontal leg in a way that works no matter which coordinate is larger?

The second issue to clarify is that it really doesn't matter which we use, $x_1 - x_2$ or $x_2 - x_1$, because this number is being squared in the general distance formula.

Note: Students will be working with the distance formula throughout the unit. In the discussion of *Defining Circles,* they will use the distance formula to develop the general equation for a circle with center (*a*, *b*) and radius *r*.

Key Question

How can you express the length of the horizontal leg in a way that works no matter which coordinate is larger?

Supplemental Activity

In *Perpendicular Bisectors by Algebra* (extension), students use the distance formula to find the equation of a perpendicular bisector of a line segment.

How Does Your Orchard Grow?

Intent

Students develop a general plan for solving the unit problem.

Mathematics

So far, students' work on the unit problem has mainly involved developing an understanding of what it means for the orchard to become a hideout. They have examined several mini-orchards and realized in each case that when the trees reach a certain trunk radius, all lines of sight out from the center become blocked.

In this activity, they will begin to examine how long it will take for the orchard to become a hideout. Their task at this stage will be to develop a general plan, which they will implement later in the unit when they get more specific information.

Students are told that the rate of growth of the trees will be given in terms of a constant rate of increase in the cross-sectional area of the trunks. Developing the general plan will reveal that not only must students consider this rate of growth in their solution, but they must also be able to determine area from circumference or from radius. This will help motivate the later exploration of the relationships between radius, circumference and area, leading to the discovery of π and development of the formulas for the area and circumference of a circle.

Progression

Students will develop a plan that involves multiple geometric components:
- determining the current cross-sectional area from the circumference and the ultimate cross-sectional area from the hideout radius
- determining how long it will take for the trees to grow to the needed cross-sectional area from the current cross-sectional area

It should not be assumed at this time that students know the area and circumference formulas for circles. Rather, the emphasis in this activity is to motivate development of those formulas.

Approximate Time

25 minutes for activity
10 minutes for discussion

Classroom Organization

Individuals or small groups, followed by whole-class discussion

Doing the Activity

Because students have not worked on the unit problem for several days, you may want to begin by having them restate the basic question: How soon after Madie and Clyde plant their orchard will the center of the lot become a true "orchard hideout"? In particular, you can refer to the two main subquestions formulated during the discussion of *Orchard Hideout*.

- How big do the tree trunks have to become for the center of the orchard to be a hideout?
- How long will it take for the tree trunks to reach that size?

Before students begin working on the activity, be sure that everyone understands the meaning of the term *cross-sectional area*. You can review that we are treating each tree trunk as a **cylinder,** which is analogous to a right prism. The concept of cross-sectional area is appropriate for any *right prism* and can be applied to cylinders as well. Also remind students that the term *circumference* refers to both the circle's boundary and the length of this boundary.

Using Specific Values to Get Started

Although the activity tells students to assume that they know both the circumference and the rate at which the cross-sectional area grows, it does not actually give them either of these values.

If students are having a hard time getting started on the problem, you might suggest that they begin by making up specific values and answering the time question for those values. They can then work from the specific case to develop a general plan.

Discussing and Debriefing the Activity

Have students from one or two groups report on the geometric questions they came up with. They should focus on two key questions.

- How do you find the cross-sectional area of a tree if you know the radius of its trunk?
- How do you find the current cross-sectional area of a tree from its circumference?

Tell students that one of their main tasks for the unit will be developing relationships among the radius, the area, and the circumference of a circle.

Note: Some students may already know formulas for finding the area and circumference of a circle in terms of the radius. If so, acknowledge this and explain that they will be learning *why* the formulas are what they are, and this understanding will enable them to develop their own formulas for future problems.

The Time Question

Have one or two other students describe how they would use the answers to the two geometric questions (and the other information they would have) to answer the time question.

As a class, build on these explanations to develop a general plan, and post it for reference during the unit.

For example, the plan might be something like this one.
1. Find the cross-sectional area for a tree whose radius is equal to the hideout tree radius.
2. Find the current cross-sectional area of the trees (from their circumference).
3. Find the difference between the cross-sectional areas (to see how much each tree has to grow).
4. Divide the difference by the annual rate of growth (to see how many years it would take).

Emphasize that to use their plan, students will need to answer the key questions about finding the cross-sectional area of a tree from either the radius or the circumference of its trunk. You may want to post those questions along with "the plan."

Key Questions

How do you find the cross-sectional area of a tree if you know the radius of its trunk?

How do you find the current cross-sectional area of a tree from its circumference?

A Snack in the Middle

Intent

Students develop the **midpoint formula**.

Mathematics

Students develop the midpoint formula by first looking at specific examples and then generalizing the process they used. Earlier work with the distance formula and Pythagorean formula is reinforced as well.

Progression

The activity initially gives students three pairs of coordinates and asks them to find the midpoints of the segments joining each pair. This is presented in the context of finding a location equidistant from Madie and Clyde at which to place a snack while they are working in the orchard. In each case, the students are asked to prove that their solution is equidistant from the given points. The whole-class discussion points out that this is only a partial proof that their solution is indeed the midpoint. Proof that the solution point actually lies on the line segment connecting the two points is left for consideration in the supplemental activity *Midpoint Proof*.

The final question asks students to generalize their work by developing a formula for the midpoint between any two points on the coordinate plane.

Approximate Time

25 minutes for activity (at home or in class)
15 minutes for discussion

Classroom Organization

Individuals, followed by whole-class discussion

Discussing and Debriefing the Activity

Ask students to reach agreement within their groups on Questions 1 through 3, and then have individual members report for their groups.

Question 1

Students will probably not have trouble coming up with (27, 10) as the point halfway between (24, 6) and (30, 14). But Question 1 also asks them to show that (27, 10) is equidistant from the other two points.

They might do this using the distance formula, or they might go directly to the Pythagorean theorem, perhaps using a diagram. For instance, they might use the right triangles in the accompanying diagram to show that both the distance from (27, 10) to (30, 14) and the distance from (27, 10) to (24, 6) are equal to $\sqrt{3^2 + 4^2}$ (which is equal to 5).

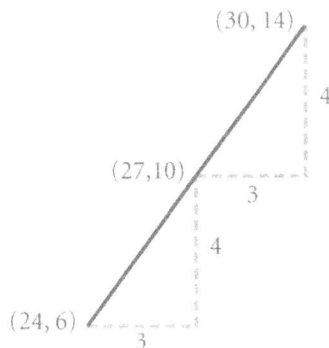

The Missing Step

Showing that (27, 10) is equidistant from (24, 6) and (30, 14) does *not* prove that (27, 10) is the midpoint of the segment connecting those two points. Proving this can be thought of as proving these two statements.

- (27, 10) is equidistant from the points (24, 6) and (30, 14).
- (27, 10) is on the segment connecting these two points.

Question 1 asks students to prove only the first of these statements. But, in fact, any point on the perpendicular bisector of a segment is equidistant from the two endpoints (as students saw in *Only Two Flowers*), so the first statement alone does not show that (27, 10) is the midpoint.

If students don't realize that establishing (27, 10) as the midpoint also entails proving the second statement, you should point out that they have not done this. (Assure them that the second statement is true, as are similar assertions for Questions 2 and 3.)

One approach to proving the second statement is to find the equation of the line through (24, 6) and (30, 14) and show that (27, 10) is on that line. That approach is explored in the supplemental activity *Midpoint Proof*.

Questions 2 and 3

Although you may want to omit the details for Questions 2 and 3, be sure to get the numerical answer for Question 3, because it involves a noninteger coordinate.

Question 4

After completing Questions 1 through 3, move on to Question 4. You may want to have a volunteer propose a general formula for finding the midpoint and then let others offer comments or additional ideas.

Students may suggest different correct formulas. For instance, two reasonable proposals for the midpoint of the segment connecting (x_1, y_1) and (x_2, y_2) are

$$\left(\frac{x_1 + x_2}{2}, \frac{y_1 + y_2}{2}\right) \text{ and } \left(x_1 + \frac{x_2 - x_1}{2}, y_1 + \frac{y_2 - y_1}{2}\right)$$

If students give more than one correct formula, have them verify that they are equivalent. You may want to post all the correct formulas that students find and label them as *midpoint formulas.*

If students had trouble finding a formula, one productive approach is to simplify the problem to one dimension. For instance, ask them to find a formula for the point on the number line that is halfway between u and v. Describing a midpoint as a kind of average may help students understand why the expression $\frac{u + v}{2}$ gives the correct result.

Working with specific examples may lead to a different (but equivalent) expression. For instance, a student might find the point halfway between 8 and 14 by finding the difference (6) and adding half of this difference (3) to 8, which gives the correct value of 11. Following through on this arithmetic in the general case would give the expression $u + \frac{v - u}{2}$.

Supplemental Activity

Midpoint Proof (extension) asks students to prove that (27, 10) is on the line segment connecting (24, 6) and (30, 14).

Equidistant Points and Lines

Intent

In *Equidistant Points and Lines*, students explore situations in coordinate geometry that involve equal distances.

Mathematics

The activities in this sequence involve developing proofs based upon coordinate geometry. Students hone their skills in logic and reasoning as they discover and prove general principles in geometry. They use the midpoint and distance formulas to justify their answers for specific situations.

Progression

Proving with Distance—Part I and *Proving with Distance—Part II* require students to repeatedly apply the midpoint and distance formulas in order to explore questions concerning a square with its vertices at given coordinates. In preparation for *Down the Garden Path*, students prove that the distance from a point to a line must be measured perpendicular to the line. *Perpendicular and Vertical* leads to a proof that a tangent is perpendicular to the radius of a circle at the point of tangency.

POW 2: On Patrol, continues with this theme of geometric discoveries and proofs. Students begin by discovering that a point that is equidistant from two intersecting lines must lie on an angle bisector, and this is proven in class discussion. They then extend this discovery by finding that the angle bisectors of a triangle intersect at the *incenter*, the center of the inscribed circle.

Proving with Distance–Part I
Down the Garden Path
Perpendicular and Vertical
POW 2: On Patrol
Proving with Distance–Part II

Proving with Distance—Part I

Intent

Students reinforce their earlier findings in coordinate geometry.

Mathematics

Students put the distance and midpoint formulas and the converse of the Pythagorean theorem to work to prove some facts about quadrilaterals. They begin by proving that a given quadrilateral is a square, and then do an exploration that leads to a conjecture: Connecting midpoints of consecutive sides in a quadrilateral will always yield a parallelogram.

Progression

The first question presents the students with the coordinates of the vertices of a quadrilateral, and asks that they prove it is a square. This involves using the distance formula four times to prove that it is equilateral; using the distance formula to find the length of the diagonal; and then applying the converse of the Pythagorean theorem to prove that it has right angles.

The next two questions ask students to draw two quadrilaterals in the coordinate plane, and then to connect the midpoints of consecutive sides. Students will have to apply the midpoint formula to find the midpoints of the four sides in both figures. They are then asked to examine the smaller quadrilaterals formed and to make a conjecture about the lengths of the sides. They will discover that when the midpoints of consecutive sides are joined, the resulting figure appears to be a parallelogram. Finally, students are asked to apply the distance formula eight more times to prove that the opposite sides of each figure are in fact equal.

Approximate Time

40 minutes for activity (at home or in class)
10 minutes for discussion

Classroom Organization

Individuals, followed by whole-class discussion

Discussing and Debriefing the Activity

Select presenters and encourage them to make a sketch of the situation. You can also ask students what distances they need to know and what formulas they have available to help them.

In Question 1, students have to show that the sides are the same length and that the angles are right angles. They will probably do the latter by applying the Pythagorean theorem to the lengths of the sides and the lengths of the diagonals.

The Converse of the Pythagorean Theorem

As usually stated, the Pythagorean theorem says (in brief, using appropriate labeling)

If $\triangle ABC$ is a right triangle, then $a^2 + b^2 = c^2$.

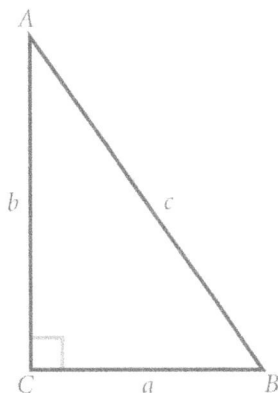

Students are using the converse of the Pythagorean theorem to prove that the angles are right angles. That is, they are showing that the triangle formed by two sides and a diagonal of the quadrilateral fits an equation like $a^2 + b^2 = c^2$, concluding that the triangle must be a right triangle.

Question 2

For Question 2, have two or three students show their results because each example will probably be different. Students should have seen that the "inner" quadrilateral formed by connecting midpoints has opposite sides of equal length and been able to verify this by finding the midpoints and applying the distance formula.

In fact, the figure is a parallelogram, and if students make this observation, you can assure them that they are correct. The supplemental activity *Midpoint Quadrilaterals* outlines a proof.

Supplemental Activity

Midpoint Quadrilaterals (extension) asks students to use the distance formula to prove that the figure formed by connecting the midpoints of consecutive sides of a quadrilateral is always a parallelogram.

Down the Garden Path

Intent

Students prove that any line through the midpoint of a segment is equidistant from the endpoints of the segment.

Mathematics

In solving the unit problem, it will be necessary for students to consider the distance from a lattice point in the orchard to a line of sight. In the discussion *distance* is defined as the shortest distance from the point to the line, and it is established that this distance must be measured perpendicular to the line.

This activity leads students to discover that any line that is equidistant from two points passes through the midpoint of the segment joining the two points. As part of the discussion following completion of the activity, students are challenged to prove that the converse of this statement is true. The proof provides a nice review of triangle congruence.

Progression

Down the Garden Path asks students to write instructions for designing a straight garden path that will be equidistant from two flowers, and then to sketch all such possible paths. Introduce this activity with discussion of just what is meant by distance from a point to a line and how such a distance should be measured. The discussion outlined below challenges students to go beyond their intuitive answer and to explain why the shortest distance from a point to a line must be the length of a line segment that is perpendicular to the line.

The discussion that introduces this activity should take place before students work on *Perpendicular and Vertical*. However, that activity may then be completed before the discussion of *Down the Garden Path.*

After brief discussion of the students' findings that all paths equidistant from the two flowers pass through the midpoint of the segment joining the two flowers, students are then asked to work in their groups to prove the converse, that any line through the midpoint of a segment is equidistant from its endpoints.

Approximate Time

20 to 25 minutes for introduction
20 minutes for activity
20 to 25 minutes for discussion and small-group work

Classroom Organization

Individuals, followed by whole-class discussion and small-group work

Doing the Activity

Down the Garden Path involves the definition of *the distance from a point to a line*. You can introduce the issue of defining this distance by having students read the introduction to *Down the Garden Path* (or you might simply describe the scenario yourself). Then show them a diagram like this one.

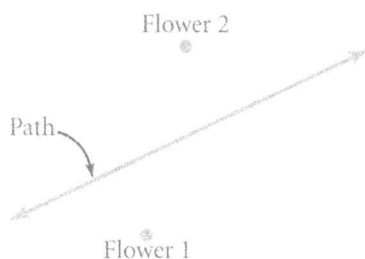

Ask, *Are the two flowers the same distance from the path?* Be sure students realize that if someone were walking along the path, that person would be closer to Flower 1 at some places and closer to Flower 2 at others.

Defining "Shortest Distance"

Clarify that we usually talk about distance *between points*, but that here the issue is the distance *from a point to a line*. Point out that many line segments connect a given point to a given line, including horizontal and vertical segments. Emphasize that these segments have different lengths, so we need a definition that establishes what "distance from a point to a line" means. A diagram like this one might help illustrate the issue.

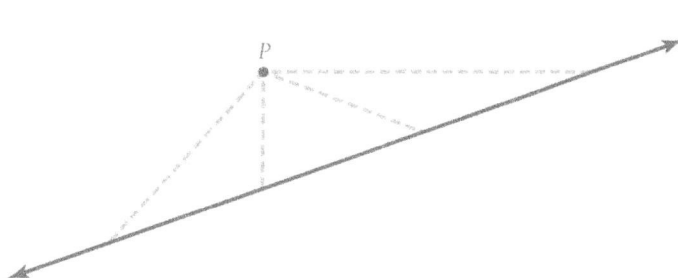

Ask, *How would you define the distance from one of the flowers to the path?* Students will probably agree that the best approach is to focus on the *shortest* distance from the given point to the given line. Tell students that this is, indeed, the formal mathematical definition:

> The *distance from a point to a line* is the length of the shortest line segment connecting the point to the line.

Shortest Paths and Perpendicularity

Have a student draw a line segment showing the "minimal distance" for Flower 1 (as shown below), and ask, How is the line segment with the shortest distance related to the path itself? The class should see that the segment is perpendicular to the path.

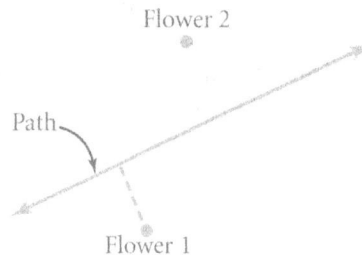

Flower 2

Path

Flower 1

Ask students how they would prove in general that the perpendicular distance is the shortest possible distance. As needed, help them develop a diagram like the one that follows. Tell them that point S is called the *foot of the perpendicular from P to l* and that T is any other point on l.

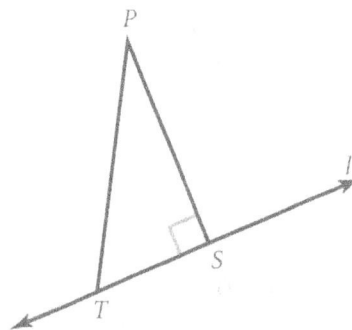

P

l

S

T

Now ask students, How can you be sure that $\overline{PT}$ is longer than $\overline{PS}$? They should see that $\overline{PT}$ is the hypotenuse of a right triangle and that $\overline{PS}$ is a leg of this right triangle, so $\overline{PT}$ is longer (by the Pythagorean theorem).

Connection to the Unit Problem

You might ask, How is the idea of distance from a point to a line connected to the unit problem? It may help to repeat the central unit question: How soon after Madie and Clyde plant their orchard will the center of the lot become a true "orchard hideout"?. Mention that lines of sight are important in the problem. In solving the unit problem, students will need to examine the distance from a tree to such a line.

Discussing and Debriefing the Activity

Begin the discussion of the activity by marking two points on the board (representing the flowers) and having several students each draw a line (representing a possible path) that is equidistant from the two points. You might get a diagram like this one:

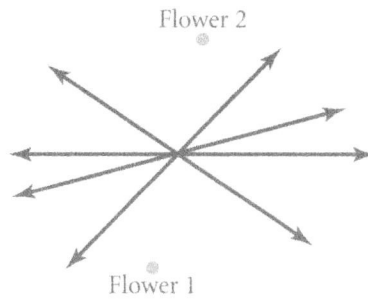
Flower 2

Flower 1

It should emerge that the lines all go through a common point. Ask, How is this common point related to the two flowers? Students should be able to see that it is the midpoint between the two flowers.

Proving Equidistance

First, have students clearly state their discovery as a conjecture. Then ask them to formulate the converse of this statement—something like "any line through the midpoint of a segment is equidistant from its endpoints." Have them work in their groups to try to prove this second statement. (The supplemental problem *Equidistant Lines* asks them to prove the original statement.)

As needed, help students develop a diagram such as this one, in which X and Y are the original points (the flowers), Z is the midpoint of the segment connecting them, and l is any line through Z.

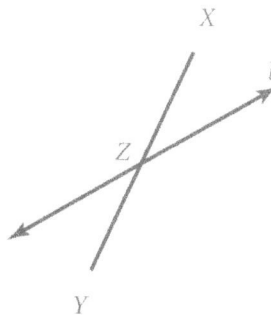

Once students have such a diagram, ask, What do you need to show? They should see that they need to show that X and Y are equidistant from l. This involves drawing perpendiculars such as $\overline{XU}$ and $\overline{YV}$ (as in the next diagram) and showing that the lengths XU and YV are equal.

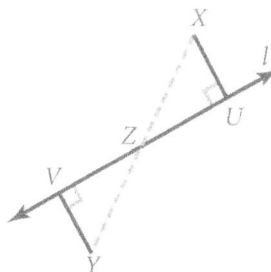

There are various approaches to this proof. For example, students can show that the triangles *XZU* and *YZV* are congruent, or they can use trigonometry to express *XU* and *YV* in terms of the equal lengths *XZ* and *YZ*.

Discussion of the congruence approach provides a good review of several ideas in *A Geometric Summary*.

For example, in triangles *XZU* and *YZV*, ∠*XZU* = ∠*YZV* because they are vertical angles. Also, both triangles have a right angle (because $\overline{XU}$ and $\overline{YV}$ are defined to be perpendicular to *l*). By the angle sum property, the third pair of angles is also equal, so the triangles are similar. Because the corresponding lengths *XZ* and *YZ* are equal, the ratio of the corresponding sides is 1, so the triangles are congruent. This means that the corresponding lengths *XU* and *YV* must also be equal.

A brief discussion of Question 3 may help students recognize the subtle distinction between the set of *lines* equidistant from two given points, and the set of *points* equidistant from two given points. The latter forms a single line, the perpendicular bisector of the segment, such that every point is simultaneously the same distance from the two endpoints; this is not true of the other lines through the midpoint.

The supplemental activity *Equidistant Lines* asks students to prove their original conjecture, which is the converse of the result just shown.

Key Questions

Are the two flowers the same distance from the path?

How would you define the distance from one of the flowers to the path?

How is the segment with the shortest distance related to the path itself?

How can you be sure that $\overline{PT}$ is longer than $\overline{PS}$?

How is the idea of distance from a point to a line connected to the unit problem?

How is this common point related to the two flowers?

What do you need to show?

Supplemental Activity

Equidistant Lines (extension) challenges students to prove that if a line is equidistant from points *A* and *B*, then it goes through the midpoint of $\overline{AB}$.

Perpendicular and Vertical

Intent

Students develop proofs of two important geometric theorems.

Mathematics

In Question 1, students apply the fact that the shortest distance from a point to a line is the perpendicular distance to prove that a tangent to a circle is perpendicular to the radius at the point of tangency. Question 2 involves an algebraic proof that vertical angles are equal.

Progression

In the first question, an illustration is shown of a line tangent to a circle, and students are asked to explain how they can be certain that the point of tangency is the point on the line that is closest to the center of the circle. They are then asked what they can conclude about the relationship between the tangent and the radius to the point of tangency. (The term *tangent to a circle* was introduced in the discussion following *More Mini-Orchards*.)

In the second question, the intersection of two lines is shown, and students are asked to find measures of three of the angles when the measure of one angle is given. They are then asked to expand their procedure into a proof that vertical angles are always equal.

Approximate Time

30 minutes for activity (at home or in class)
15 minutes for discussion

Classroom Organization

Individuals, followed by whole-class discussion

Discussing and Debriefing the Activity

Question 1a

Ask a volunteer to present Question 1a. The basic approach is shown in the diagram here, in which B is any point on line *l* other than A itself. The key idea is that B must be outside the circle, because otherwise *l* would not be tangent to the circle.

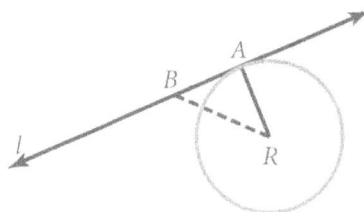

Because B is outside the circle, its distance from R is greater than AR. This means that A is the point on *l* closest to R.

To prove formally that B is outside the circle and that points outside the circle are "farther" from the center than points on the circle, you would need careful definitions of the concepts of "inside" and "outside." In formal mathematics, these are sophisticated concepts, and you need not get into those subtleties here.

Question 1b

Ask for a volunteer to state her or his conclusion for Question 1b. Although students may see from the diagram that $\overline{RA}$ is perpendicular to *l*, the focus should be on *why* this is the case. Emphasize the connection between "shortest distance" and perpendicularity.

This is a good occasion to emphasize that "distance from a point to a line" is *defined* to be the shortest distance, but that we *prove* that the perpendicular segment is that shortest distance.

The discovery in Question 1b will be important to the solution of the unit problem. Be sure to have the students summarize it clearly:

A tangent to a circle is perpendicular to the radius at the point of tangency.

Question 2

Students should be able to find the other angles in Question 2a by using the idea of supplementary angles. Because the purpose of Question 2a is to serve as a model for the generalization in Question 2b, have students explain clearly how they are finding each angle.

Then ask for a general proof, going back to the specific example as needed. For instance, if students can state clearly that they found $\angle 2$ in Question 2a as $180° - 20°$, they can use this to write $\angle 2$ in general as $180° - \angle 1$. Similarly, $\angle 3 = 180° - \angle 2$, so $\angle 3 = 180° - (180° - \angle 1)$, which shows that $\angle 3 = \angle 2$.

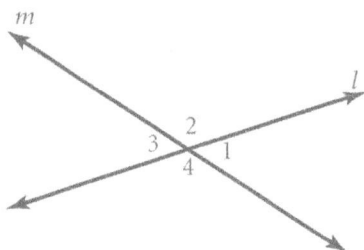

POW 2: On Patrol

Intent

Students discover that the angle bisectors of a triangle intersect at a single point.

Mathematics

POW 2: On Patrol requires students to work again with the distance from a point to a line, a concept that is critical to solving the unit problem. In this activity, it will be discovered and then proven that a point that is equidistant from two intersecting lines must lie on the bisector of an angle formed by the intersection. This leads to the further discovery that the angle bisectors of a triangle intersect at the incenter of the triangle, which is the center of the inscribed circle.

Progression

This activity asks students to describe where to place a highway patrol station so that it will be equidistant from two, then three, then four or more intersecting straight highways. In order to solve this POW, the students must first come to the realization that a point that is equidistant from two intersecting lines must lie on the bisector of one of the angles formed by the intersection. The situation with two highways leads to this discovery, and it is proven in the whole-class discussion.

Students are then able to use that fact to address the situation with three highways. Remember to consider the possibility that two or even three highways are parallel or that all three intersect at a single point, but the situation that leads to discovery of the more important mathematical principles is that where the three lines form a triangle. Although four possible locations offer solutions, the one where the three angle bisectors of the triangle's interior angles intersect is the incenter— the center of the inscribed circle. This is also the location with the shortest distance to the three lines.

With more than three highways, it is not always possible to find a solution.

Students will work on this POW independently for about a week, followed by two or three student presentations and then whole-class discussion.

Approximate Time

5 to 10 minutes for introduction
2 to 3 hours for activity (at home)
5 minutes for selection of presenters
25 minutes for presentations and discussion

Classroom Organization

Individuals, followed by student presentations and whole-class discussion

Doing the Activity

This POW is very similar to *POW 1: Equally Wet*, except that we start with two lines instead of two points. Help students connect this problem to the concept of the distance from a point to a line, which they just worked with in *Down the Garden Path*.

We hope that students will build on their discussion of *POW 1: Equally Wet* and approach this problem in terms of a *set of points* rather than an individual solution. (The one obvious solution is to put the station right at the intersection of the highways, although some people might argue that you can't put a station in the middle of a highway.)

It may help to remind students that the solutions to *POW 1: Equally Wet* formed a line, namely, the perpendicular bisector of the segment connecting the two flowers.

Students should be given about a week to work on this POW. On the day before the POW is due, choose three students to make POW presentations on the next day, and give them overhead transparencies and pens to take home to use for preparing presentations.

Discussing and Debriefing the Activity

Have the students make their presentations. The main concern here is the two-highway case (Question 1). It is more important to have a good discussion of the solution for this case than a partial discussion of the general problem.

The Two-Highway Case

As students present the case of two intersecting highways, be sure to have them give a diagram of the situation. In the diagram below, the dashed lines represent possible locations for the station.

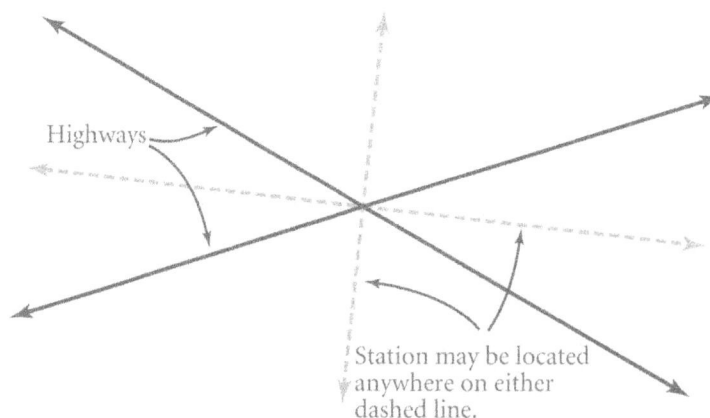

Highways

Station may be located anywhere on either dashed line.

Note: Students may have thought about only one of the two lines that form solutions to the problem. If so, then you may want to work with that single line of solutions as described here and come back later to the fact that there are two lines.

Next, have students turn to a characterization or description of the dashed lines. Because students worked with angle bisectors in *Another Kind of Bisector*, they may immediately recognize that the dashed lines are the bisectors of the angles formed by the highways.

If not, you might say, "In the earlier POW, the two flowers defined a line segment. What kind of mathematical object does the intersection of two highways define?" This should suggest to students that they focus on an angle formed by the highways.

You can follow this up by asking, How is the set of solutions related to the angle formed by the highways? Continue with this line of questioning until students recognize the dashed lines as the **angle bisectors**. Once this term has been introduced into the POW discussion, work with students to elicit a succinct description of the solution to the two-highway case of the POW. Here is an example of such a description:

> **The places where the station could go form the bisectors of the angles formed by the highways.**

"If and Only If" and Converses

This discussion provides an opportunity to review the use of the phrase "if and only if" and the concept of a *converse*.

Students might use the phrase "if and only if" to state the mathematical essence of the result in this way:

> **A point is equidistant from two intersecting lines if and only if it lies on the bisector of one of the angles formed by those lines.**

Students might break this statement down into its two components.

- If a point is equidistant from two intersecting lines, then it lies on the bisector of one of the angles formed by those lines.
- If a point lies on the bisector of one of the angles formed by two intersecting lines, then it is equidistant from those lines.

Be sure students understand that these two statements are converses of each other.

Reminder: Students proved in *Another Kind of Bisector* that the two angle bisector lines are perpendicular to each other.

Proving the Angle Bisector Property

If it doesn't come up in the POW presentations, ask, How would you prove that your solution is correct? This is a good issue for them to discuss in groups. You might point out that this task is similar to the activity *A Perpendicularity Proof*, although students worked on only one direction of the proof in that activity.

As a hint, have students set up a diagram like the one here, showing the two "highways," an angle bisector, and a point D on the angle bisector.

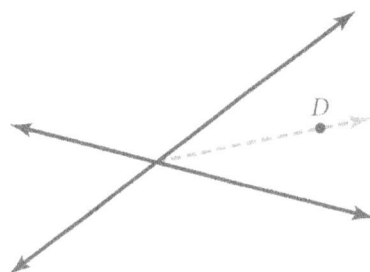

You may also want to review the meaning of "distance from a point to a line." (See the discussion introducing *Down the Garden Path.*) This should lead to the introduction of points at the feet of the two perpendiculars from D to the lines, as shown in the next diagram.

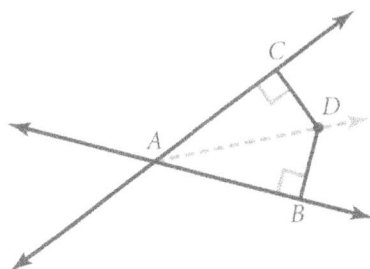

Thus, students need to prove the statement
$BD = CD$ if and only if $\angle BAD = \angle CAD$.

Students might use various approaches. Here are outlines of two, one using trigonometry and the other using congruence.

- In the two right triangles, $\sin(\angle BAD) = \dfrac{BD}{AD}$ and $\sin(\angle CAD) = \dfrac{CD}{AD}$, so $BD = AD \cdot \sin(\angle BAD)$ and $CD = AD \cdot \sin(\angle CAD)$. This shows that the angles are equal if and only if BD and CD are equal. (This proof assumes that $\sin(\angle X) = \sin(\angle Y)$ if and only if $\angle X = \angle Y$, which students may accept intuitively at this point even though it has not been formally established.)
- If either of the two conditions $\angle BAD = \angle CAD$ or $BD = CD$ is true, then the triangles ABD and ACD must be congruent, in which case the other of the two conditions must also be true.

If students use the second approach, have them explain how they would prove that triangles ABD and ACD were congruent if they knew that angles BAD and CAD were equal or if they knew that BD and CD were equal. As with the proof from *Down the Garden Path,* this will build on ideas in *A Geometric Summary.*

Optional: Constructing an Angle Bisector
If the class discussed the construction of the perpendicular bisector of a line segment with *A Perpendicularity Proof*, this is a good opportunity to follow up on the construction process

Three Highways
Part a of Question 2 focuses on the possible ways three lines can intersect. The "standard" case is that the highways create a triangle, as illustrated by the solid lines in the following diagram. However, it is also possible for two or even three of the highways to be parallel or for all three to intersect at a single point.

For the standard case, there are various ways to describe the solution. One approach is to say that the station must lie on all three angle bisectors (or, more precisely, on one line of each of the three *pairs* of angle bisectors). There are actually four solutions, as shown in the diagram, one of which is inside the triangle formed by the three highways. (The solid lines are the highways, and the dashed lines are the various angle bisectors.) Although all four points are equidistant from the three highways, the one inside the triangle is the closest of the four to the highways and would probably be the preferred solution in the real-world situation.

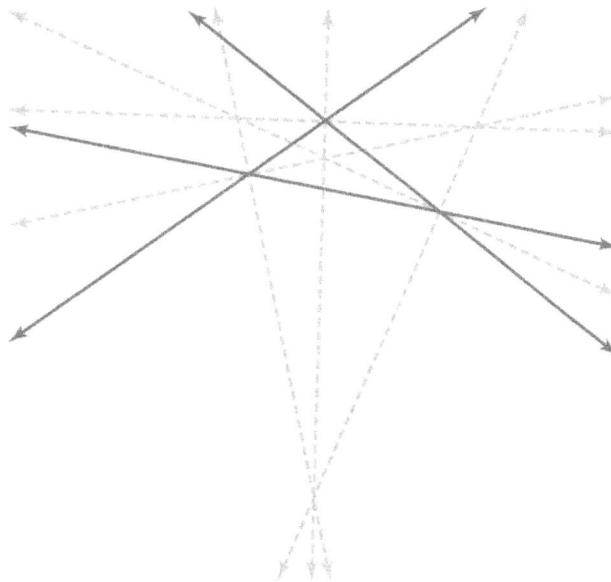

In each of the four possible cases, the common distance from the station to the three highways defines a circle. The next diagram shows the circle corresponding to the intersection point inside the triangle formed by the highways. This circle, which is tangent to all three sides of the triangle, is inscribed in the triangle and is sometimes called the *incircle.* Its center is called the *incenter* of the triangle.

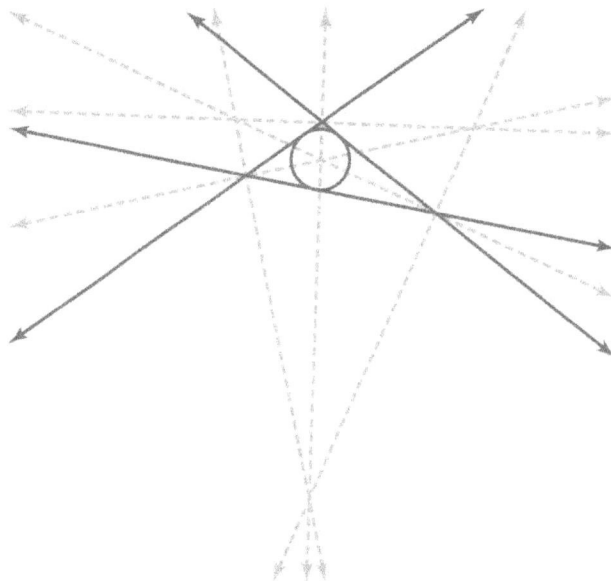

It will probably help to compare the solution for *POW 2: On Patrol*, which involves a circle *inscribed* in a triangle, to students' results from *POW 1: Equally Wet*, in which the solution involved a circle *circumscribed* about a triangle. (The circumscribed circle is sometimes called the *circumcircle*.)

Just as there are four points that are equidistant from all three highways, so also there are four circles that are tangent to all three highways. The next diagram shows all of these circles.

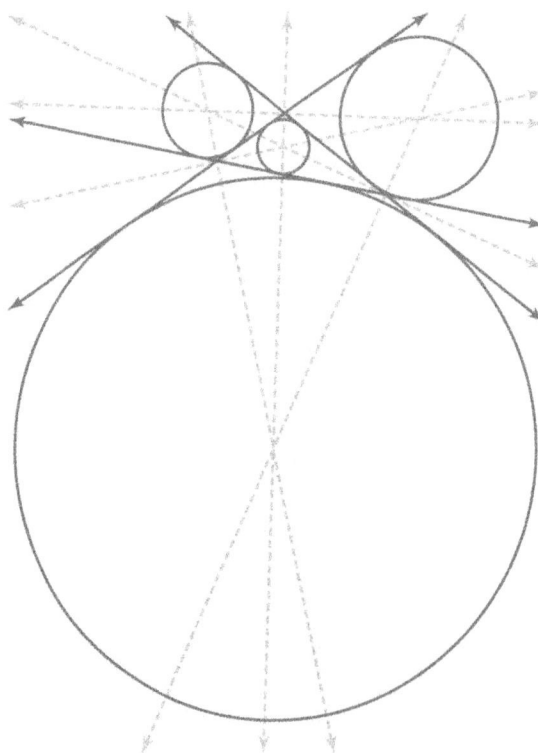

The supplemental activity *The Inscribed Circle* asks students to prove the key element in the three-highway case—that the bisectors of the three angles of a triangle meet in a single point.

Note: There are three "non-standard cases" of the three-highway problem. If all three lines intersect in a single point, the only solution is at the point of intersection (though again this may be impractical for a patrol station). If all three lines are parallel, then no solution exists. If two of the lines are parallel and the third line intersects them, there are two solutions, which can be generated through a series of constructions.

More than Three Highways

For more than three highways, one approach is to first consider three of the highways. As discussed, there are four possible locations for the station. For any one of those locations, you can observe that for that particular location to work for any other highway, the additional highway must be tangent to the corresponding circle.

Key Questions

What kind of mathematical object does the intersection of two highways define?

How is the set of solutions related to the angle formed by the highways?

How would you prove that your solution is correct?

Supplemental Activities

The Inscribed Circle (extension) involves proving that the angle bisectors of a triangle must intersect at a single point.

Medians and Altitudes (extension) asks students to prove that the three altitudes of a triangle meet in a single point, and that the three medians do so as well.

Proving with Distance—Part II

Intent

This activity reinforces students' previous work in coordinate geometry.

Mathematics

Students use the midpoint and distance formulas to explore situations involving circles circumscribed about and inscribed in a square.

Progression

The first two questions in this activity use a square with vertices at the same coordinates as in *Proving with Distance–Part I.* The first question asks students to find the center and the radius of the circumscribed circle, and the second question asks students to find the center and radius of the inscribed circle. In both cases, students are instructed to prove that their answers are correct.

In the first question, students will need to find the midpoint of a diagonal and then use the distance formula to find the distance to each of the vertices of the square. In the second question, students will need to find the midpoint of each of the sides of the square, and then find the distance from the center to each of those midpoints.

The final question asks if there is a circumscribed circle for every quadrilateral. As students discovered in *POW 1: Equally Wet*, there will only be a circumscribed circle if the four vertices happen to be equidistant from a single point.

Approximate Time

30 minutes for activity (at home or in class)
15 minutes for discussion

Classroom Organization

Individuals, followed by whole-class discussion

Discussing and Debriefing the Activity

You can have one or two students present their ideas on Question 1.

One approach is to recognize that the center of the circumscribed circle is at the midpoint of either diagonal. With this approach, students can find the coordinates of the center, (8.5, 19.5), using the midpoint formula. You can tell students that the center of a circumscribed circle is called the **circumcenter**.

Once students have the coordinates of the center, they can then use the distance formula to get the radius. For instance, the distance from the center to (5, 20) is $\sqrt{(8.5-5)^2 +(19.5-20)^2}$, which simplifies to $\sqrt{12.5}$, or approximately 3.54.

Students can prove that their answers are correct by applying the distance formula three more times to demonstrate that the distance from the center of the circle to each of the other vertices of the square is also approximately 3.54.

For Question 2, students may realize that the center of the inscribed circle is the same as that of the circumscribed circle. Once they identify the center, they need to find the distance from the center to one of the midpoints. The radius in this case is exactly 2.5.

Note: Frequent use of the terms *circumscribed* and *inscribed* in context will help students get used to them. These words recur in today's activity. While you are discussing the activity you might point out that when a circle is inscribed in a square, we can also say that the square is circumscribed about the circle, and vice versa

Question 3
Question 3 is essentially a review of the conclusions from *POW 1: Equally Wet.* For a circle to be circumscribed about a quadrilateral, its center must be equidistant from all four vertices (just as the sprinkler in the POW needed to be equidistant from all four flowers).

You can review the idea that for a polygon to have a circumcenter, the perpendicular bisectors of the sides must all intersect in a single point. This should help students see that some quadrilaterals have no circumscribed circle.

Generalizing Inscribed Circles
Students may wonder why Question 3 did not also ask about inscribed circles. If so, tell them that they can think about this in connection with their work on *POW 2: On Patrol.*

All About Circles

Intent

The activities in *All About Circles* introduce π and its application in calculating the area and circumference of a circle.

Mathematics

By calculating the area and perimeter of a series of regular polygons, with progressively larger numbers of sides, circumscribed about a circle, students recognize that the perimeter is proportional to the radius of the circle and that the area is proportional to the square of the radius. They also observe that the proportionality constant for the perimeter is exactly twice that for the area. As the number of sides increases and the polygons begin to more accurately approximate the shape of the circle, it becomes easier to accept that these same principles of proportionality must apply to the circumference and area of the circle itself.

Eventually the approximation of the proportionality constant for the area of a circle becomes accurate enough that students recognize it as **pi**. Developing π in this way as the ratio of the area to the square of the radius of the circle gives students a greater understanding of π as something more than an arbitrary number. It also makes the formulas for area and circumference of the circle flow naturally from application of the proportionality precepts that students saw in working with the circumscribed polygons.

Both the students' earlier work with the equation of a circle and their work here with finding the area of a polygon involve application of the Pythagorean theorem. A natural extension is to introduce **Pythagorean triples**.

Progression

Squaring the Circle begins a sequence of activities in which students calculate the perimeter and area of a polygon circumscribed about a circle. In subsequent activities, regular polygons with more and more sides are used. This culminates in *Polygoning the Circle*, where students work with polygons having relatively large numbers of sides, and it becomes evident that the constant of proportionality for the relationship $A:r^2$ is approaching what the students recognize as π. With this leading into the area and circumference formulas for a circle, students apply those formulas to the unit problem in *Orchard Growth Revisited*.

In *Another Kind of Bisector,* students work with angle bisectors, necessary to the solution of *POW 2: On Patrol*, introduced earlier. *Proving Triples* introduces Pythagorean triples.

Squaring the Circle

Using the Squared Circle
Hexagoning the Circle
Octagoning the Circle
Polygoning the Circle
Another Kind of Bisector
Proving Triples
POW 3: A Marching Strip
Orchard Growth Revisited

Squaring the Circle

Intent

Students explore the relationship between the area and perimeter of a square and that of its inscribed circle. This activity is the first step in a process by which students will develop the formulas for the area and **circumference** of a circle.

Mathematics

Many students will be familiar with the area and circumference formulas for a circle, but they may never have seen an explanation for the fact that π appears in both formulas. One of the goals of this sequence of activities is to give them insight into this constant and to deepen their understanding of both formulas.

Progression

Squaring the Circle asks students to draw a square with an inscribed circle on graph paper, to estimate and find the ratio of the areas of the two figures, to estimate and find the ratio of the circumference to the perimeter, and then to repeat this process. Finally, they are asked to explain why they think the ratios will or will not be the same for any circle. The goal of this activity is for students to understand that these ratios are not dependent upon the size of the circle, so that the results of the next several activities can be generalized to apply to all circles.

Approximate Time

25 minutes for activity
10 minutes for discussion

Classroom Organization

Pairs, followed by whole-class discussion

Materials

String and a compass for each pair of students
Squaring the Circle blackline master

Doing the Activity

Historical note: For centuries, mathematicians struggled with the problem of trying to use certain formal "construction" methods to draw a square whose area is the same as that of a given circle. The German mathematician Ferdinand von Lindemann showed in 1875 that this is impossible. You may want to tell students that the title of this activity refers to an ancient problem of trying to create a square whose area is *the same* as that of a given circle. (That is not what they are doing in this activity.)

We recommend that you have students do this activity in pairs (rather than in groups of four) so that more students will go through the process of measuring and counting.

As you circulate, look for pairs with good explanations for Questions 4 and 5, and ask them to prepare explanations on overhead transparencies for the class. *Note:* Students will need the results of this activity for *Using the Squared Circle*.

Discussing and Debriefing the Activity

Area Ratios
Ask several students for the estimate they got for the ratio of the area of a circle to the area of its circumscribed square, and record those estimates. Ask, What was your estimate for the area ratio? Although some variation is likely, you should get approximately the same ratio from all students. (The exact ratio is $\frac{\pi}{4}$, which is about 0.785.)

Then ask, Did you get the same ratio when you repeated the task? and Did you think that the ratio would be the same for all the circles? They should have an intuitive sense that the exact ratio will be the same, no matter what the size of the original circle. Assure them that this is so.

In fact, students may believe this even if they got slightly different answers when they repeated Question 1, because their ratios were based on separate estimates of the area of the circle and the area of the square.

During the discussion, students should gradually build their intuition about why this area ratio should be the same for all circles.

It is important to have the class agree on a specific estimate for the value of the area ratio for use in *Using the Squared Circle*. You may want to post this value (clearly labeled as an estimate) for reference.

Circumference-to-Perimeter Ratios
Use a similar sequence of questions regarding the ratio of circumference to perimeter, recording the estimates for the ratio and asking whether the ratio should vary from case to case. Again, intuition should suggest that the ratio will be the same, no matter what the size of the original circle. As with area, have the class agree on a specific value as an estimate for the perimeter ratio.

It will be convenient to refer to this ratio simply as a "ratio of perimeters," analogous to the ratio of areas. Point out again, if needed, that circumference is simply a special kind of perimeter.

Comparing the Two Ratios

Students may notice that the two ratios—the ratio of areas and the ratio of perimeters—appear to be the same. In fact, they are equal, and students will probably see this as reasonable, because both represent how big a circle is compared to its circumscribed square.

In this case, however, one ratio is comparing perimeter to circumference, and the other is comparing area to area, and so far there is no proof that these two ratios are the same. You can let students speculate about this, but for now, if it comes up, this should be left as an open question. Students will see why the ratios are equal by the completion of *Polygoning the Circle*.

Questions 4 and 5

Ask the pairs you chose earlier to present their explanations for Questions 4 and 5 to the class.

One good explanation is based on "changing the units," using a pair of diagrams like the ones below. *Note:* These diagrams are included in the blackline masters, so you can make overhead transparencies for this discussion.

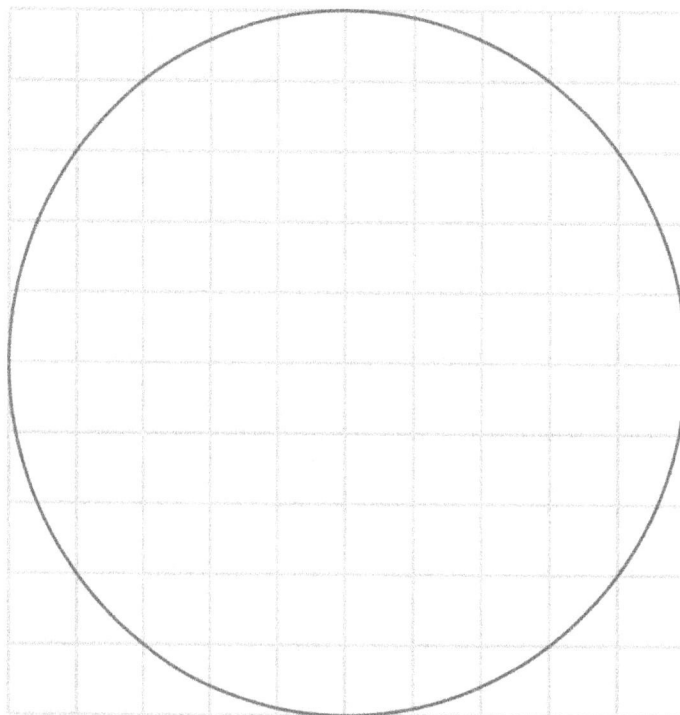

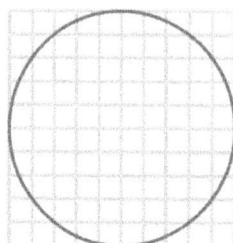

When you "blow up" (expand) a circle, you can think of blowing up the unit of measurement as well. Each of the circles shown above can be thought of as having a radius of 5 units, but the unit for the larger figure is three times that of the smaller circle. This is easily illustrated with an overhead transparency of one of the diagrams by moving the projector closer to and further from the screen, thus reducing and enlarging the image.

In the diagrams, the number of square units making up the area of the large circle is the same as the number of square units for the smaller circle, and the number of square units making up the area of the large square is the same as the number of square units for the smaller square. Therefore, the two circle-to-square area ratios must be equal.

A similar argument can be made comparing circumferences and perimeters in terms of the different unit lengths.

Key Questions

What was your estimate for the area ratio?

Did you get the same ratio when you repeated the task?

Did you think that the ratio would be the same for all the circles?

Using the Squared Circle

Intent

Students develop approximate formulas for the area and circumference of a circle.

Mathematics

This activity will give students a chance to use the ratios found in *Squaring the Circle* as a first step toward developing formulas for the area and circumference of a circle. They will discover that the circumference of a circle is proportional to the radius and that the area is proportional to the square of the radius.

Progression

Using the Squared Circle gives students a circle of radius 10 inscribed in a square. Students are asked to find the dimensions, perimeter, and area of the square, and then to use the ratios from *Squaring the Circle* to estimate the circumference and area of the circle. Finally students are asked to explain how they used the ratios to find the circumference and perimeter and to express this procedure as a formula.

The discussion establishes that the circumference is proportional to the radius and the area is proportional to the radius squared. These proportionalities are expressed as formulas and posted. Students discover that being able to find the exact perimeter and area of the circumscribed square can give us approximations of the two constants of proportionality. These values are also posted for later comparison to those derived from other circumscribed polygons.

Approximate Time

25 minutes for activity (at home or in class)
10 to 15 minutes for discussion

Classroom Organization

Individuals or small groups, followed by whole-class discussion

Doing the Activity

This activity makes use of the ratios the class agreed upon following *Squaring the Circle.* If assigning this as homework, be sure to remind students to record the ratio values.

Discussing and Debriefing the Activity

You might assign each group either the problems for circumference (Questions 2 and 3) or the problems for area (Questions 4 and 5), and give group members a few minutes to share ideas before beginning presentations.

Note: We will assume for Questions 1a and 1b that students used $\frac{3}{4}$ for both ratios found in *Squaring the Circle.* If your class agreed on different values, you will need to adjust the results shown here accordingly.

Question 2
In Question 2a, presenters should explain why the length of a side of the square is twice the radius of the circle, so the square has sides of length 20 and a perimeter of 80. They might use a diagram like this one, which shows a second radius for the given circle, a radius that is more easily related to the square.

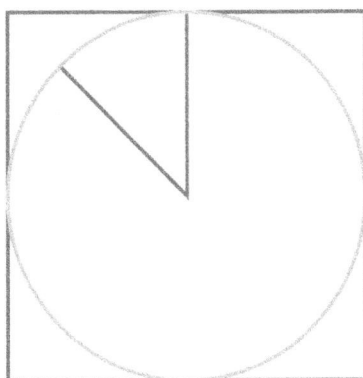

If students use $\frac{3}{4}$ for the ratio in Question 1a, they should get an estimate of $\frac{3}{4} \cdot$ 80 = 60 for the circumference of the circle.

Question 3
Students should be able to generalize the arithmetic from Question 2 to see that if the circle has radius r, then the square has side $2r$ and perimeter $8r$. This means (continuing to use $\frac{3}{4}$ for the ratio) that the circumference of the circle is approximately $\frac{3}{4} \cdot 4 \cdot 2r$, which simplifies to $6r$.

Help students see that the coefficient 6 in the expression $6r$ is an approximation based on the estimated ratio of $\frac{3}{4}$. Tell them that we will use the symbol k_c to represent the exact coefficient of r in the formula for the circumference of a circle. In other words, the formula for circumference will be written as

   $C = k_c r$

where C is the circumference and r is the radius.

Thus, k_c corresponds to the coefficient 6 (in the expression $6r$). This coefficient was found above using the value $\frac{3}{4}$ as the circumference-to-perimeter ratio from Question 1a, so the value 6 is a rough estimate for k_c. The key idea, more broadly, is that the circumference of any circle can be found by multiplying the radius by some fixed number, which is the same for every circle.

Go over the phrase "proportional to" in this context, by expressing the formula in this way.

The circumference of a circle is proportional to its radius.

Let students know that this means the same thing as the formula and that another way to express this is to say the ratio between the circumference and the radius is the same for all circles—the ratio is simply the constant k_c. (In the next subsection, students will see that the *area* of a circle is proportional to the *square* of its radius.) Tell students that k_c is called a **constant of proportionality** (or *proportionality constant*).

Questions 4 and 5

Presenters should go through a similar process to find the area of a circle. They should see that they can double the radius to get the side of the circumscribed square, square that number to find the area of the circumscribed square, and then multiply the area of the square by the estimated ratio to find the area of the circle.

Suppose, for example, that they use $\frac{3}{4}$ for the ratio from Question 1b. In the specific example from Question 4, they would find that the area of the square is 20^2 $= 400$, so that the area of the circle is $\frac{3}{4} \cdot 400 = 300$. For an initial circle of radius r, they would get an area of $(2r)^2$ for the square and $\frac{3}{4} \cdot (2r)^2$ for the circle. Have students expand and simplify these expressions, getting $4r^2$ for the area of the square and $3r^2$ for the area of the circle.

As with the circumference discussion, help students see that the coefficient 3 in the expression $3r^2$ is based on the estimated ratio of the areas of the circle and the circumscribed square (in Question 1b). Tell students that we will use the symbol k_a to represent the exact coefficient of r^2 in the formula for the area of a circle. In other words, the formula for area will be written as
$$A = k_a r^2$$
where A is the area and r is the radius.

Thus, k_a corresponds to the coefficient 3 (in the expression $3r^2$), which students found using the value $\frac{3}{4}$ as the area-to-area ratio from Question 1b. Thus, the value 3 is an estimate for k_a. Here, students should see that the area of any circle can be found by multiplying the square of the radius by a fixed number.

Go over the phrase "proportional to" in this context as well, by expressing the formula in this way.

The area of a circle is proportional to the square of its radius.

Identify k_a as another constant of proportionality.

Summary

Post the formulas

$$C = k_c r \qquad \text{and} \qquad A = k_a r^2$$

with appropriate labeling and explanation. Tell students that they will devote the next few days to understanding these two formulas and getting a better idea about the actual values of the two proportionality constants k_c and k_r.

Ask, How would such formulas fit into your work on the unit problem? Bring out that such formulas are precisely what was called for in the questions listed in connection with "the plan" they developed in *How Does Your Orchard Grow?*

Summarize what they have done so far by making a table something like this one.

Shape	Perimeter	Area
Circle	$k_c r \approx 6r$	$k_a r^2 \approx 3r^2$
Square	$8r$	$4r^2$

Introduce the terms *perimeter coefficient* and *area coefficient*, respectively, for numbers such as 8 and 4 in this table. As the labeling of the table suggests, it's usually convenient to use the same terminology for the circle, even though, strictly speaking, we should use the term *circumference coefficient* for k_c. As noted previously, students should see that *circumference* and *perimeter* are analogous.

Ask, Do you notice anything about the coefficients? They may notice that the perimeter coefficient for the square is twice the area coefficient for the square and that their tentative value for k_c is twice the tentative value for k_a. You can tell them that they will be learning more about the reason for this relationship as they expand the table. *Note:* If this observation does not come out at this point, you need not bring it up. You can raise it as part of the discussion of the next activity.

Key Questions

How would such formulas fit into your work on the unit problem?

Do you notice anything about the coefficients?

Hexagoning the Circle

Intent

Students find the perimeter and area of a regular hexagon circumscribed about a circle.

Mathematics

The activity *Hexagoning the Circle* is a natural follow-up to *Squaring the Circle* and to *Using the Squared Circle*. Students acquire an even better approximation for the constants of proportionality for the area and circumference of a circle as they find the perimeter and area of a regular hexagon circumscribed about a circle. With this second pair of values added to their table, they are able to recognize a pattern that k_c is twice k_a.

Progression

This activity gives students the same radius 10 circle as in the preceding activities, but now with a regular hexagon circumscribed about it. Students are asked to find the perimeter and area of the hexagon, then to generalize this process to a circle of radius r.

Subsequent discussion of the solutions will need to be careful and thorough, as students often struggle with this. Further discussion will then bring out that the hexagon's perimeter coefficient is twice its area coefficient, and demonstrate that this will be true for any regular polygon circumscribed about the circle, and ultimately for the circle itself.

Approximate Time

25 to 35 minutes for activity
10 to 15 minutes for discussion

Classroom Organization

Small groups or individuals, followed by whole-class discussion

Doing the Activity

Introduce the activity by asking, What circumscribed polygon might better approximate the circumference and area of the circle than a square? Students will probably suggest a variety of polygons that look "rounder" than a square. Tell them that they will begin their quest for better approximations for k_c and k_a by finding the perimeter and area of a circumscribed regular hexagon.

As students work, you can identify groups to make presentations. You may want to have more than one presentation each for area and perimeter.

Discussing and Debriefing the Activity

You can have students from the groups you selected make presentations.

Perimeter

Begin with presentations on the perimeter of the hexagon, because students will first need to find the length of a side before they can find the area. The discussion here is for the general case (Question 2), but you may prefer to have students explain the specific case (Question 1) first.

Students will probably recognize that the hexagon can be broken up into six equilateral triangles, as shown in the accompanying diagram. Be sure to get an explanation of why the triangles in the diagram are equilateral.

The key step in developing a formula for the perimeter of the hexagon is finding the length of a side of one of the equilateral triangles. Finding this length uses the fact that $\overline{AD}$ is a radius of the circle, so its length is r.

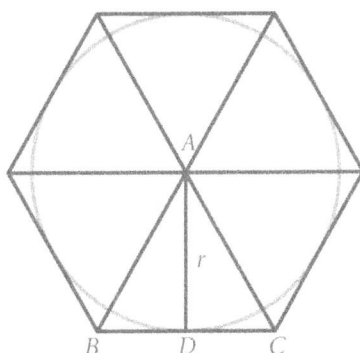

There are a variety of approaches to finding the length BC. Here are three of them.

First Approach

$$\angle ABD = 60°, \text{ so } \sin 60° = \frac{AD}{AB} = \frac{r}{AB}$$

therefore $AB = \dfrac{r}{\sin 60°} \approx 1.15r$

so $BC \approx 1.15r$

Second Approach

Similarly, $\tan 60° = \dfrac{AD}{BD} = \dfrac{r}{BD}$

So $BD = \dfrac{r}{\tan 60°} \approx 0.577r$

Because $BC = 2 \cdot BD$ we get $BC \approx 2 \cdot 0.577r \approx 1.15r$

Third Approach

Let $x = BC$, so $BD = \dfrac{x}{2}$

Applying the Pythagorean theorem to $\triangle ABD$ gives $\left(\dfrac{x}{2}\right)^2 + r^2 = x^2$

Simplifying, we get $r^2 = \dfrac{3}{4}x^2$

So $x^2 = \dfrac{4}{3}r^2$

and, once again, $x = \left(\sqrt{\dfrac{4}{3}}\right)r \approx 1.15r$

Because the length of the perimeter is $6 \cdot BC$, this gives a perimeter of approximately $6.93r$. *Caution:* Rounding may give slight inconsistencies in these values. For instance, $6 \cdot 1.15$ gives 6.90, but $6 \cdot \sqrt{\dfrac{4}{3}}$ is 6.93 to two decimal places.

The approach using the Pythagorean theorem works only for the regular hexagon, and not for regular polygons with different numbers of sides. Therefore, be sure to discuss at least one approach that uses trigonometry. If no one mentions the approach using the Pythagorean theorem, you need not bring it up.

The supplemental activity *Thirty-Sixty-Ninety* asks students to find the exact values of the trigonometric functions for angles of 30° and 60°. This is an appropriate time to refer them to this problem.

Area

After the presentations on perimeter, have students turn to area. The key step now is realizing that the area of $\triangle ABC$ is $\dfrac{1}{2} \cdot AD \cdot BC$. Because $AD = r$ and $BC \approx 1.15r$, this gives an area for $\triangle ABC$ of approximately $0.58r^2$. The area of the entire hexagon is six times as large, or approximately $3.46r^2$.

Have students add this information as a new row to their table, so that the table now looks like this.

Shape	Perimeter	Area
Circle	$k_c r \approx 6r$	$k_a r^2 \approx 3r^2$
Square	$8r$	$4r^2$
Regular hexagon	$\approx 6.93r$	$\approx 3.46r^2$

The Relationship Between the Coefficients

Students will be going through the process of finding the perimeter and area several times, for different polygons. Eventually, they should conclude that k_c is exactly twice k_a. We recommend that for now you focus on the numerical coefficients found for the regular hexagon.

Ask, *Do you see any relationship between the two coefficients you just found?*, meaning the perimeter coefficient for the hexagon (which is approximately 6.93) and the area coefficient for the hexagon (which is approximately 3.46). They will probably recognize that the perimeter coefficient is about twice the area coefficient. (It is exactly twice, but rounding might suggest otherwise.)

Ask, *Why does this relationship hold?* As a hint, ask students how they got the area of the regular hexagon. Bring out that the area of $\triangle ABC$ is $\frac{1}{2} \cdot AD \cdot BC$, which is $\frac{1}{2} \cdot r$ times the length of the base of that triangle. Because this relationship holds true for each of the six triangles, the overall area is $\frac{1}{2} \cdot r$ times the perimeter.

The factor $\frac{1}{2}$ explains why the area coefficient is half the perimeter coefficient.

What about the factor r? Be sure students see that this factor explains why the perimeter column of the table involves multiples of r and the area column involves multiples of r^2.

Emphasize that the reasoning used to relate the coefficients shows that the perimeter coefficient is *exactly* twice the area coefficient, even though the values in the table, 6.93 and 3.46, might suggest otherwise. Point out that these numerical values come from approximations of square roots or trigonometric functions.

You may want to review the case of the circumscribed square to see that the coefficients 8 and 4 are exact values, so the same relationship holds true for that case as well. Ask if students can explain the relationship for the square using the same reasoning as for the hexagon.

Students should gradually see that the reasoning used for the circumscribed regular hexagon applies to any circumscribed polygon. Because the circle is a limiting case of regular polygons, this means that k_c must be exactly twice k_a.

Key Questions

What circumscribed polygon might better approximate the circumference and area of the circle than a square?

Do you see any relationship between the two coefficients you just found?

Why does this relationship hold?

What about the factor *r*?

Supplemental Activity

Thirty-Sixty-Ninety (reinforcement) reminds students of the relationships between the side lengths for a thirty-sixty-ninety degree triangle.

Octagoning the Circle

Intent

Students continue to lay the foundation for understanding π as a constant of proportionality.

Mathematics

Continuing the progression from the several previous activities, students now find the perimeter and area of a regular octagon circumscribed about a circle, and they reconfirm their earlier discovery that the perimeter coefficient is twice the area coefficient.

Progression

Octagoning the Circle asks students to find the perimeter and area of a regular octagon circumscribed about a circle of radius 10 and then to generalize their work for a circle of radius *r*. They will then add the new approximations for the proportionality constants to the table.

Approximate Time

25 minutes for activity (at home or in class)
10 to 15 minutes for discussion

Classroom Organization

Individuals, followed by whole-class discussion

Doing the Activity

If time allows, you can introduce this activity by asking, *What polygon might you look at next?* Many students will recognize that the more sides a polygon has, the better it approximates the circle. If you have students who want to try a 50-gon or 100-gon, tell them that they will get their opportunity in an upcoming activity. Here, everyone will work with an octagon.

If assigning this as homework, you may want to give a few students overhead transparencies and pens to take home so that presentations can be done at the beginning of class tomorrow.

Discussing and Debriefing the Activity

Begin with presentations of the solutions.

The perimeter and area of the circumscribed regular octagon can be found in various ways using trigonometry. We outline one approach here, but variations are certainly likely, depending on which angle is used.

Because the circumscribed octagon contains eight central triangles, the central angle for each is 45°. As shown here, the small angle in $\triangle ABD$ is 22.5°. Therefore,

$\tan 22.5° = \dfrac{BD}{AD} = \dfrac{BD}{r}$, so $BD = (\tan 22.5°) \cdot r$.

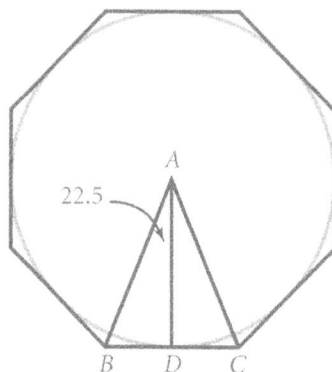

This means that $BC = 2 \cdot (\tan 22.5°) \cdot r$, and so the overall perimeter is $8 \cdot 2 \cdot (\tan 22.5°) \cdot r$, which is approximately $6.63r$.

The area of $\triangle ABC$ is $\dfrac{1}{2} \cdot AD \cdot BC$. But AD is a radius, so the area of the triangle is $\dfrac{1}{2}r \cdot [2(\tan 22.5°)r]$, which equals $(\tan 22.5°) \cdot r^2$. Therefore, the overall area is $8 \cdot (\tan 22.5°) \cdot r^2$, which is approximately $3.31r^2$.

Relating Area and Perimeter
As was done previously with the hexagon, be sure to bring out the relationship between the perimeter and area formulas. Students should see that the area of $\triangle ABC$ is exactly $\dfrac{1}{2}r$ times the length BC, so the area of the octagon is $\dfrac{1}{2}r$ times the perimeter of the octagon.

Be sure students see that the factor $\dfrac{1}{2}r$ accounts for two things.

- The area coefficient is exactly half the perimeter coefficient.
- The area formula has r^2 while the perimeter formula has simply r.

In our analysis, these perimeter and area constants are $16 \cdot \tan 22.5°$ and $8 \cdot \tan 22.5°$, respectively. Encourage students to use expressions such as $\tan 22.5°$, rather than their approximate numerical value, until their final computation. This will help to minimize the final approximation error and also bring out the relationship between the coefficients.

Expanding the Table

Add the area and perimeter information for the octagon to the table students have been developing. Ask, *What happens to k_c and k_a as the number of sides increases? Why?* (Basically, we want students to see that both are decreasing, because the circumscribed polygons are getting "closer" to the circle.)

Shape	Perimeter	Area
Circle	$k_c r \approx 6r$	$k_a r^2 \approx 3r^2$
Square	$8r$	$4r^2$
Regular hexagon	$\approx 6.93r$	$\approx 3.46r^2$
Regular octagon	$\approx 6.93r$	$\approx 3.31r^2$

Key Questions

What polygon might you look at next?

What happens to k_c and k_a as the number of sides increases? Why?

Polygoning the Circle

Intent

As students find the perimeter and area of a regular polygon with more than eight sides circumscribed about a circle, they obtain an accurate enough estimate of the area coefficient for the circle to recognize it as π.

Mathematics

Polygoning the Circle culminates in the definition of π and a discussion of the relationship between the perimeter and area formulas for a circle.

Progression

The activity asks students to develop formulas for the area and perimeter of a regular polygon having a number of sides other than 4, 6, or 8, circumscribed about a circle of radius r. The discussion afterward introduces π as the area coefficient that students have been estimating with increasing accuracy in the preceding activities.

Approximate Time

35 minutes for activity
10 to 20 minutes for discussion

Classroom Organization

Individuals, followed by whole-class discussion

Materials

Optional: *The Story of Pi* video

Doing the Activity

You may want to assign groups specific polygons to work on so that the class will see a variety of presentations.

Discussing and Debriefing the Activity

Let students from several groups share the formulas each group found for the perimeter and area of its regular polygon. While this is being done, you can add the results to the table created so far, repeated here for your convenience.

Shape	Perimeter	Area
Circle	$k_c r \approx 6r$	$k_a r^2 \approx 3r^2$
Square	$8r$	$4r^2$
Regular hexagon	$\approx 6.93r$	$\approx 3.46r^2$
Regular octagon	$\approx 6.93r$	$\approx 3.31r^2$

As these results unfold, you should help students focus on these key points.

- The perimeter is always a particular constant times the radius, where the constant (which we are calling the *perimeter coefficient*) depends on the number of sides.
- The area is always a particular constant times the square of the radius, where, again, the constant (the *area coefficient*) depends on the number of sides.
- The area coefficient for a particular regular polygon is always exactly half the corresponding perimeter coefficient.
- The larger the number of sides, the more closely the constants found should approximate k_c and k_a.

The Area Coefficient Is Half the Circumference Coefficient

The final step before defining π and stating the formulas for the area and circumference of a circle is to bring together the last two key points in our list.

Ask students, What is the relationship between k_a and k_c?, and have them explain that relationship.

Technically, the explanation involves the concept of **limit**, but students will probably be able to give fairly clear intuitive descriptions of the relationship between the coefficients for the individual polygons and the coefficients for the circle. Because the area coefficient for each polygon is exactly half of that polygon's perimeter coefficient, the same relationship holds between the coefficients for the circle.

Defining π

With appropriate fanfare, tell students that they have essentially solved the problems posed in the discussion of *How Does Your Orchard Grow?* That is, they have found the relationship between the radius of a circle and its circumference and area.

Emphasize that they still have only approximate values for k_c and k_a but that they have seen that k_a is exactly half of k_c.

You might ask, Do the approximate values for k_a suggest anything to you? (Some of them may recognize the values as being close to π, perhaps building on prior knowledge of the area formula.)

If no one knows, tell the class that as the number of sides of the circumscribed polygon grows, these area coefficients approach a number that is traditionally represented by the Greek letter π (and pronounced like "pie"). In other words, π can be defined by the relationship

$$A = \pi r^2$$

where r is the radius of a circle and A is the area of that circle.

Historical note: The first use of the symbol π for the circumference-to-diameter ratio was in 1706, by the English writer William Jones (1675–1749). The symbol came into general use in this sense when the Swiss mathematician Leonhard Euler (1707–1783) adopted it in 1737.

The Approximate Value of π

As students saw in the discussion of *Polygoning the Circle*, the value of this number is roughly 3, and their work with regular polygons gives them other estimates for π.

Tell students that this number is approximately 3.14, emphasizing that this is only an approximation. Tell them as well that there is no finite decimal that gives the value exactly, nor is there any fraction that gives it exactly. (You may want to mention the term *irrational number* in this context.)

Have students spend a few minutes in their groups relating the coefficients k_c and k_a to this special number π. Then bring them together to share conclusions.

They should see that π has essentially been defined to be the constant they have been calling k_a. Further, because k_a is exactly half of k_c, they should see that $k_c = 2\pi$. Again, with appropriate fanfare, help students put this all together, and post the two formulas

$$C = 2\pi r \text{ and } A = \pi r^2$$

with accompanying labels.

Remind students that the area of a circle is exactly $\frac{1}{2} r$ times the circumference,

just as the area for each circumscribed polygon was $\frac{1}{2} r$ times the corresponding perimeter.

Tell students that although we have here defined π as the ratio of the area of a circle to the square of its radius, π historically has been defined as the ratio of the circumference of a circle to its diameter. Students have seen that these two ratios must be the same, so it doesn't matter which is used as the formal definition.

Finding and Representing π

Tell students that in various ways, mathematicians have found better and better approximations for π (far more exact than would ever be needed for practical measurement). Point out that students' graphing calculators have a key that gives a good approximation for the value of π. (Many of them already know this.)

Emphasize that it is important to have a symbol to represent the number because the decimal representation of π continues indefinitely without a repeating pattern.

Key Questions

What is the relationship between k_a and k_c? Why?

Do the approximate values for k_a suggest anything to you?

Supplemental Activity

Darts (reinforcement) requires students to work with the formula for the area of a circle in the context of an expected value problem in probability.

Another Kind of Bisector

Intent

Students are introduced to **angle bisectors**.

Mathematics

In *Another Kind of Bisector*, students use algebra to prove that the bisectors of the angles formed by two intersecting lines are perpendicular.

Progression

The activity begins by defining angle bisectors and then having students once more calculate the remaining angles when given one of the angles formed by two intersecting lines. The second question, in several more steps, asks students to prove that the bisectors of the angles in the first problem are perpendicular. The final question asks them to build on the arithmetic of the specific case to prove, in general, that the angle bisectors are perpendicular. Angle bisectors play a key role in *POW 2: On Patrol.*

Approximate Time

25 minutes for activity (at home or in class)
10 minutes for discussion

Classroom Organization

Individuals, followed by whole-class discussion

Discussing and Debriefing the Activity

Question 1 of this activity is essentially the same as Question 2a of *Perpendicular and Vertical,* so you can simply have a volunteer give the size for each of the angles, without any discussion.

On Question 2a, let a volunteer present a proof. The key idea is to show that the sum of angles *PAC, CAQ, QAD,* and *DAR* is 180°. Students should be able to find each of these angles using the results from Question 1 and the definition of an angle bisector, as shown in the next diagram. (For this question, students need not break down $\angle CAD$ into the two component angles $\angle CAQ$ and $\angle QAD$.)

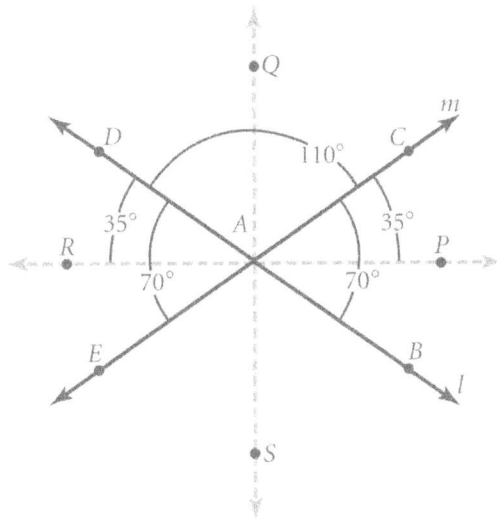

Be sure to elicit a clear statement that proving *P*, *A*, and *R* collinear is equivalent to proving that ∠*PAR* is 180°. Review the term **straight angle** to refer to an angle whose defining rays form a straight line.

You need not explicitly discuss Question 2b, because the reasoning is identical with Question 2a. The proof for Question 2c is also quite similar, but we recommend that you go through the details as preparation for Question 3.

Question 3

The reasoning in Question 3 is identical to that for Question 2, but it is more difficult because students have to work in general terms. If they have trouble, ask explicit questions, such as, What is ∠*PAC* in terms of *x*? What is ∠*CAD*? Students should see that the sum of angles *PAC, CAQ, QAD,* and *DAR* can be expressed algebraically as

$$\frac{x}{2} + \frac{1}{2}(180° - x) + \frac{1}{2}(180° - x) + \frac{x}{2}$$

They can then show that this simplifies to 180°. The proof of the perpendicularity is similar.

Angle Bisectors and the POW

You may want to drop the hint that angle bisectors play a role in *POW 2: On Patrol* that is similar to the role of perpendicular bisectors in *POW 1: Equally Wet.*

Key Question

What is ∠*PAC* in terms of *x*? What is ∠*CAD*?

Proving Triples

Intent

Students gain familiarity with **Pythagorean triples**.

Mathematics

After introducing the Pythagorean triples of 3, 4, 5 and 5, 12, 13, *Proving Triples* challenges students to prove that any multiple of a Pythagorean triple will be another Pythagorean triple. A hint advises them of the possibility of both geometric and algebraic approaches.

Progression

Students begin by verifying that 5–12–13 is a Pythagorean triple. Question 2 asks them to determine that multiplying the triple 3–4–5 by 3 will form another Pythagorean triple, and students are then asked to speculate (with examples) as to whether a multiple of a Pythagorean triple will always form another Pythagorean triple. Question 3 asks them to prove their conjecture.

Both geometric and algebraic proofs are discussed.

Approximate Time

30 minutes for activity (at home or in class)
10 minutes for discussion

Classroom Organization

Individuals, followed by whole-class discussion

Discussing and Debriefing the Activity

You may want to skip straight to Question 3 and ask if anyone thinks they have a proof that any multiple of a Pythagorean triple is also a Pythagorean triple.

Here are outlines of the two approaches—one geometric, one algebraic—mentioned in the activity.

- If a, b, and c are the lengths of the sides of a triangle, then a triangle with sides of lengths xa, xb, and xc will be similar because of the proportionality of the sides. Therefore, if sides of lengths a, b, and c form a right triangle, so will sides of lengths xa, xb, and xc.
- If $a^2 + b^2 = c^2$, then $x^2(a^2 + b^2) = x^2c^2$. Therefore, $x^2a^2 + x^2b^2 = x^2c^2$, and so $(xa)^2 + (xb)^2 = (xc)^2$. In other words, if a, b, and c fit the Pythagorean equation, then so do xa, xb, and xc.

You can refer students to the supplemental activity *More About Triples* for more work on the concept of Pythagorean triples. The supplemental activity *Counting Trees*, while more elementary, is also related to the idea of Pythagorean triples.

Supplemental Activities

Counting Trees (extension) leads students to discover that the number of boundary trees in every orchard with a positive integer radius is a multiple of eight.

More About Triples (extension) asks students to consider whether the lengths of the sides of an isosceles right triangle could form a Pythagorean triple and to perform several proofs dealing with primitive Pythagorean triples.

POW 3: A Marching Strip

Intent

Students develop a general formula based on seeing a pattern or by using a general analysis.

Mathematics

POW 3: A Marching Strip has no direct mathematical connection to the unit theme. By examining concrete examples in this POW, students try to discover a unifying principle, most likely by examining the number patterns that form. The solution introduces the *greatest common divisor*.

Progression

The POW describes the floor of a rectangular courtyard covered with square tiles. Students are asked to predict how many tiles will be touched by a diagonal line if the courtyard has 63 rows of 90 tiles, and then if the courtyard has r rows of c tiles.

Students should be given at least a week to complete the POW, followed by presentations by several students.

Approximate Time

10 minutes for introduction
30 minutes for small group work (optional)
1 to 3 hours for individual work (at home)
15 to 25 minutes for presentations and discussion

Classroom Organization

Individuals, followed by student presentations and whole-class discussion

Doing the Activity

Take a few minutes to go over the situation in *POW 3: A Marching Strip*. The example discussed here will probably be sufficient to clarify for students what they are being asked to do.

The accompanying diagram shows a courtyard with four rows of tiles and six tiles in each row. The light squares represent the ordinary tiles; the dark squares represent the extra-strength, expensive tiles. In this case, the king needs eight extra-strength tiles.

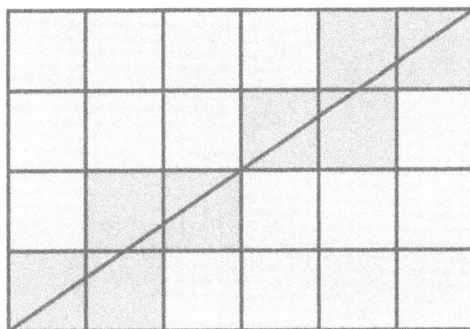

You might give students some time to work on *POW 3: A Marching Strip* in their groups. Discussion of this POW should be scheduled for at least a week away.

As you circulate around the room, watch carefully for groups that are guessing about whether a tile includes part of the diagonal. Help them to come up with a plan that will answer this question with certainty. Be sure students realize that if the diagonal goes through only a corner of a tile, that tile need not be one of the expensive kind.

On the day before the POW is due, choose three students to make POW presentations on the following day, and give them overhead transparencies and pens to take home to use for preparing presentations.

Discussing and Debriefing the Activity

Ask the selected students to make their presentations, and then let others add any further ideas.

One interesting element of this POW is seeing how students move from concrete examples to the generalization, perhaps using pictures. You can let the level of student interest in the problem help you determine how much time to spend and how much depth to go into on the discussion of this problem.

Note: The number of tiles needed for the 63x90 courtyard is 144.

The General Formula
The general formula is $r + c - \text{GCD}(r,c)$ [where $\text{GCD}(r,c)$ means the *greatest common divisor of r and c*], but you should not necessarily expect students to find the formula, or to be able to explain it. Some students may get the formula by examining number patterns. For instance, they may notice that the number of special tiles needed seems to be approximately $r + c$.

If they look at the difference between the actual answer and $r + c$ in various cases, they may see that this difference is often 1, that the difference is even when both numbers are even, and so on. By looking at examples, they may recognize this difference as the GCD of r and c.

Orchard Growth Revisited

Intent

Students refine their plan for solving the unit problem, now incorporating the formulas for area and circumference of a circle.

Mathematics

In *How Does Your Orchard Grow?*, students were asked to describe a plan for finding out how long it would take for the tree trunks to reach a given radius. This activity asks them to carry out that plan, based on some specific information. This requires that the students apply their newly discovered formulas for the circumference and area of a circle.

Progression

Orchard Growth Revisited gives students some of the missing information for the unit problem: the current circumference of the tree trunks, and the growth rate in terms of cross-sectional area. Students are then asked to calculate how long it will take for the trees to grow to a radius of one foot. Though they do not yet know what the true hideout radius will be, this activity enables them to see how the circle formulas will work in that process. Students will also see that the information provided here is sufficient to find the time required for the tree trunks to reach any given radius

Approximate Time

5 minutes for introduction
25 minutes for activity (at home or in class)
10 minutes for discussion

Classroom Organization

Individuals, followed by whole-class discussion

Doing the Activity

We recommend that you take some time to go over the plan for solving the unit problem that was developed in *How Does Your Orchard Grow?* in preparation for this activity.

Discussing and Debriefing the Activity

The discussion of this activity should give you a sense of how well students can use the area and circumference formulas developed previously.

Let one or two volunteers present their results. You may want to have them sketch diagrams showing the trees now (with a circumference of 2.5 inches), the trees eventually (with a radius of 1 foot), and perhaps some intermediate stages.

They might find the time required using these steps.
- Find the current radius of the trees from their current circumference. ($2\pi r = 2.5$ inches, so $r = \dfrac{2.5}{2\pi} \approx 0.40$ inches)
- Find the current cross-sectional area of the trees from their radius. [$A \approx \pi(0.40)^2 \approx 0.50$ square inches]
- Find the cross-sectional area (*in square inches!*) for a tree with trunk radius of 1 foot. [$A = \pi(12)^2 \approx 452.39$ square inches]
- Find the required increase in cross-sectional area. (Increase $\approx 452.39 - 0.50 \approx 451.89$ square inches)
- Divide the increase in cross-sectional area by the growth per year. ($\dfrac{451.89}{1.5} \approx 301$ years)

That's a long time to wait!

Make sure students realize that the "1-foot" assumption was simply for the sake of working out the method and that they still have no idea what the hideout tree radius is for the orchard of radius 50. They don't even know the unit distance between trees.

The round-off issues here are interesting, because squaring the radius changes the percentage of round-off error. We recommend that you allow generous variation in results here and later in the unit, and not get distracted by complexities such as significant digits. The focus here should be on the geometry, and it's sufficient if students see that it takes several hundred years for the trees to reach a radius of 1 foot.

This is also a good occasion to remind students that as they make successive calculations, they should use the calculator's multidigit values from one step as they do the next step. Usually, they can do this without actually entering the value (for instance, using an [ANS] key).

Cable Complications

Intent

In this sequence of activities, students broaden and deepen their understanding of various equations relating to circles as they also begin to see how all of the pieces will fit together. They develop the tools they'll need to find the tree radius that will turn their orchard into a true hideout.

Mathematics

In *Orchard Growth Revisited*, students explored how to fit the various elements together to solve one of the two primary sub-questions to the unit problem, how long it will take the trees to grow to a specific radius. *Cable Ready* will require that they examine the other major issue, what tree radius is required to block a given line of sight. This fundamentally involves finding the distance from a point to a line, but that process synthesizes a number of important geometric principles: the perpendicularity of the tangent to the radius at the point of tangency, trigonometric functions and their inverse functions, the distance formula, similarity, algebra, and the Pythagorean theorem.

Other activities in this unit reinforce what the students have learned so far about the equation of a circle and the area and circumference of a circle, while expanding their understanding of those topics.

Progression

Cable Ready will take a couple of hours of class time. The problem can be solved in a number of ways, each of which requires students to apply a larger number of principles than they are accustomed to bringing to bear on a single problem. It is very useful to have student presentations and to take the time to examine each solution path carefully.

Going Around in Circles gives students practice in manipulating the circle area and circumference formulas forward and backward. *Daphne's Dance Floor* explores the effects of a change in scale on the area and circumference of a circle.

Students' facility with the equation of a circle is strengthened through *Defining Circles, The Standard Equation of the Circle,* and *Completing the Square and Getting a Circle*. These develop the equation of a circle having an arbitrary center (thus expanding the previous development of the equation for a circle centered at the origin) and the transformation of the equation of a circle into standard form by completing the square.

Cable Ready
Going Around in Circles
Daphne's Dance Floor

Defining Circles
The Standard Equation of the Circle
Completing the Square and Getting a Circle

Cable Ready

Intent

Students solve a problem analogous to the unit problem.

Mathematics

This is another good opportunity to see if students can relate what they've been doing to the overall unit problem, as they find the distance from a point to a line in a coordinate setting.

Progression

Cable Ready tells students that an electrical cable runs from (0, 0) to (30, 20) in their orchard. They are asked what the largest radius orchard is that they could plant without placing a tree directly on top of the cable. The second question asks how big the tree trunks would have to become before one of them bumped into the cable, if an orchard of that radius were planted.

The problem can be solved in a number of ways, any one of which requires students to synthesize a number of concepts. The student work is followed by presentations and discussion of as many of these methods as you can find represented in your students' work.

Approximate Time

5 minutes for introduction
75 minutes for activity and preparation for presentations
35 minutes for presentations and discussion (More time may be needed if you decide to have every group present.)

Classroom Organization

Small groups, followed by student presentations and whole-class discussion

Materials

Poster materials, including large-grid poster paper

Doing the Activity

Where Are We Now?

From the discussion of *Orchard Growth Revisited*, students should realize that they can figure out how long it will take for trees to reach any given size. So the key task is to find the hideout tree radius. That is, they must determine how big the trees need to get in order to create a true "orchard hideout." Ask, Where are you now in relation to the overall problem? What do you know? What do you still need to find out?

You may want to let the class brainstorm briefly how to answer these questions. You might suggest that students think about when a particular line of sight would be blocked by a particular tree. Where is the tree closest to the origin that would be on the cable path? Help them to see that this is related to the idea of the distance from a point to a line, a concept integral to *Down the Garden Path* and *POW 2: On Patrol*.

Cable Ready

Cable Ready will help students figure out how big the trees need to get to hide the center. Depending upon your schedule, students are likely to need more than one day in which to complete work, prepare posters, and make group presentations.

Hints for Question 1
Students should be able to complete Question 1 by sketching the path of the cable and marking the trees on a coordinate grid.

As needed, help them to see that the first tree that would have been planted along the cable path would be the one at (3, 2). In other words, Madie and Clyde could have planted any orchard that did not include this tree.

Ask, How big a mini-orchard could you have that did not include the tree at (3, 2)? A tree at (3, 2) would have been more than 3 units from the origin (because $\sqrt{3^2 + 2^2} > 3$), so they could have planted a complete mini-orchard of radius 3 without planting on the cable path. But this tree would have been less than 4 units from the origin (because $\sqrt{3^2 + 2^2} < 4$), so Madie and Clyde could not have planted a complete mini-orchard of radius 4.

Hints for Question 2
Once students see that Madie and Clyde could only have planted a mini-orchard of radius 3, they need to find out how big the trees could have grown before reaching the cable.

If a hint is needed, you can ask, What trees in the radius 3 mini-orchard are closest to the cable? They should be able to see intuitively that the closest trees would be those at (2, 1) and at (1, 1). (If any groups need an additional challenge, ask them to show that these two points are the same distance from the cable. They might see that this follows from their work in the activity *Down the Garden Path*.)

The real challenge here is finding the distance from (2, 1) or (1, 1) to the cable. Several methods are outlined in the discussion of the activity.

Note: This is a difficult activity, but experience has indicated that students will get it and that the process of working it through, by putting together various ideas that they have learned, will be worthwhile.

If groups finish early, ask them to look for other ways to solve the problem.

Discussing and Debriefing the Activity

Begin by having a student from one group present Question 1. You might ask why, in Question 1c, students were to assume that the tree trunks were very thin.

You might have another student merely set up Question 2 with a diagram and a statement of the main problem of the activity. For instance, the student might present a diagram like this one, showing the first-quadrant portion of the mini-orchard of radius 3.

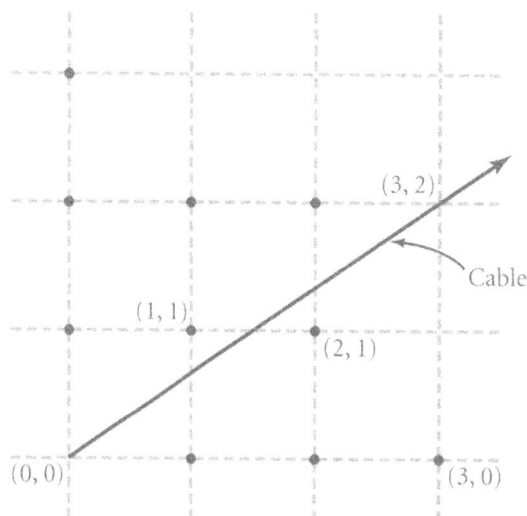

In this diagram, the dots represent the trees in this portion of the orchard and the line represents the cable. Using a diagram like this, a student should explain that the task is to find the distance from (2, 1) [or (1, 1)] to the line representing the cable.

This is a good time to raise the question, *Why are the trees at (2, 1) and at (1, 1) equidistant from the cable?* If needed, suggest that students connect this question with their work on *Down the Garden Path*, in which they saw that any line through the midpoint of a segment is equidistant from the endpoints of the segment. In *Cable Ready,* they need to show that the cable goes through the midpoint of the segment connecting (2, 1) and (1, 1).

For the overall discussion, it will probably be helpful to have the presenter identify the distance being sought and label several key points, as in the following diagram. It is especially important for students to understand that point *C* is the foot of the perpendicular from *B* to the cable, so that $\overline{BC}$ is, by definition, perpendicular to

$\overline{AC}$. You might also have students explain why the points (3, 2) and $(1\frac{1}{2}, 1)$ are on the line representing the cable.

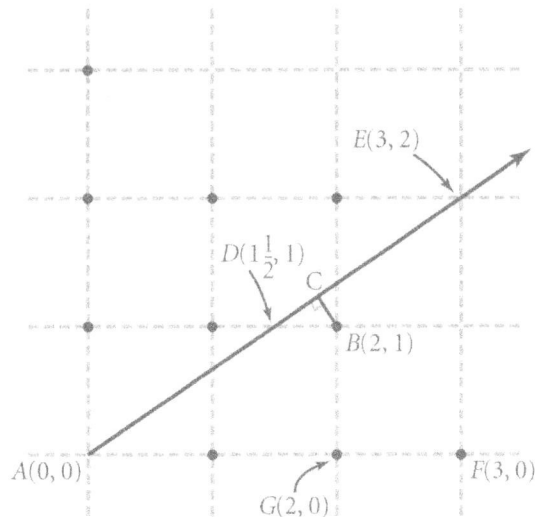

Based on such a diagram, students should recognize that their goal is to find the distance BC from point (2, 1) to the cable line. Ask, What distance are you trying to find?

We suggest that you have several presentations on how to find this distance—several approaches are possible, each using very different reasoning. Because of this variety, students may not be familiar with an approach being presented, so it is especially important to encourage questions. If you think some detail has not been made clear, you might ask for a volunteer to restate that particular idea.

You should post the conclusion from these presentations, which is that the trees will reach the cable when they achieve a trunk radius of $\frac{1}{\sqrt{13}}$ units. (It's fine if students use an approximation, such as 0.28 units.) Students will use this result in *The Other Gap*.

Be sure that students explain where the numbers they use come from so that someone who used a different approach can follow their explanation.

We provide outlines here of three of the many methods, using elements of the previous diagram. You need not get presentations of all of these methods, but try to get as much variety as possible.

Using Trigonometry

In the diagram below, $\tan(\angle BAG) = \frac{1}{2}$ and $\tan(\angle EAF) = \frac{2}{3}$. From these facts, we can find both angles and subtract to get $\angle CAB$. (This gives $\angle CAB \approx 7.1°$.)

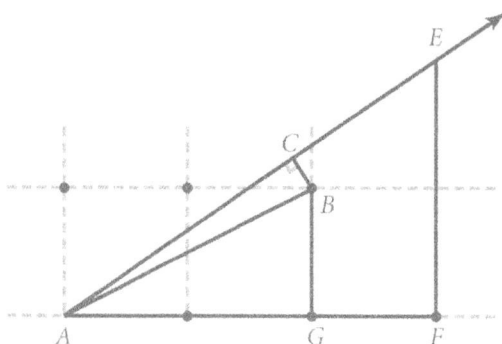

Then we can use right triangle ABC in which we now know $\angle CAB$ and can find the length AB from the distance formula. We can then use the equation $\sin(\angle CAB) = \dfrac{BC}{AB}$ to find the length BC. (This gives $BC \approx 0.28$.)

Using Similar Triangles

In the next diagram, $\triangle DCB$ is similar to $\triangle AFE$, because both are right triangles and both have an angle formed between the cable and a horizontal line.

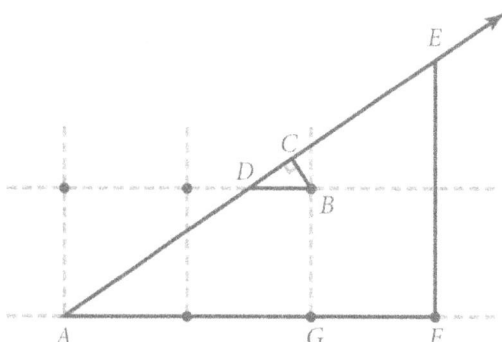

Comparing shorter legs and hypotenuses, we get $\dfrac{BC}{EF} = \dfrac{BD}{EA}$.

We have $BD = \dfrac{1}{2}$ and $EF = 2$, and we can get $EA = \sqrt{13}$ from the Pythagorean theorem. Substituting these values and simplifying gives $BC = \dfrac{1}{\sqrt{13}}$.

Using the Pythagorean Theorem, Similarity, and Algebra

In the previous diagram, $BC^2 + CD^2 = BD^2$ by the Pythagorean theorem. The similarity explained previously shows that the lengths BC and CD are in the ratio of 2 to 3, so we can call these lengths $2x$ and $3x$.

Substituting these expressions for *BC* and *CD* and substituting $\frac{1}{2}$ for *BD* into the

Pythagorean theorem gives $(2x)^2 + (3x)^2 = \left(\frac{1}{2}\right)^2$, which simplifies to $13x^2 = \frac{1}{4}$,

and so $x = \frac{1}{\sqrt{52}}$. Because $BC = 2x$, this gives $BC = \frac{2}{\sqrt{52}}$ (which students may or

may not simplify to $\frac{1}{\sqrt{13}}$.

Key Questions

Where are you now in relation to the overall problem? What do you know?

What do you still need to find out?

What is the tree closest to the origin that would be on the cable path?

How big a mini-orchard could you have that did not include the tree at
(3, 2)?

What trees in the radius 3 mini-orchard are closest to the cable?

Why are the trees at (2, 1) and at (1, 1) equidistant from the cable?

What distance are you trying to find?

Going Around in Circles

Intent

Students gain additional experience using the formulas for the area and circumference of a circle.

Mathematics

Going Around in Circles expands students' fluency with the circumference and area formulas for a circle. It requires them not only to use radius and diameter to solve for circumference and area, but to manipulate the formulas for the more difficult task of finding radius or diameter given the circumference or area.

Progression

Four relatively straightforward problems involving area or circumference of a circle are presented. Each solution requires more than a cursory familiarity with the circle formulas, and they should be carefully discussed.

Approximate Time

20 minutes for activity (at home or in class)
10 minutes for discussion

Classroom Organization

Individuals, followed by whole-class discussion

Discussing and Debriefing the Activity

Let students from a few groups share their solutions with the rest of the class. Although the essence of these problems is the application of the formulas for area and circumference, this is not necessarily an elementary task.

In particular, keep in mind that Questions 3 and 4 are more complicated than Questions 1 and 2, because they involve applying the formulas "in reverse." For instance, in Question 1, students are given the radius of the orchard and asked to find its area. In Question 4, by contrast, students are given an area and need to find the corresponding diameter.

Daphne's Dance Floor

Intent

Students explore how a change of scale affects area and circumference.

Mathematics

In spite of the appearance of r^2 in the formula for the area of a circle, students often initially make the mistake of assuming that applying a scale factor to the radius scales the area by the same factor. *Daphne's Dance Floor* requires students to rediscover the principles that while circumference is directly proportional to radius, area is proportional to the square of the radius.

Progression

Students are presented with a suggestion that the circumference and area of a circular dance floor might be estimated by considering them to be directly proportional to the circumference and area of a smaller circle. They are then asked to estimate the values using this supposition of proportionality and then to calculate the values using the circle formulas. The discussion following the activity points out that the general principles observed here (that circumference is proportional to radius, and that area is proportional to the square of the radius) apply to polygons as well.

Approximate Time

20 minutes for activity (at home or in class)
15 minutes for discussion

Classroom Organization

Individuals, followed by whole-class discussion

Discussing and Debriefing the Activity

An important goal of this activity is to bring out these two conclusions.
* Multiplying the radius (or diameter) of a circle by a given amount multiplies its circumference by the same amount.
* Multiplying the radius (or diameter) of a circle by a given amount multiplies its area by the square of that amount.

Some students may be familiar with these principles from previous experiences. Nevertheless, you should probably assume that these are new ideas.

Questions 1 and 2

Ask volunteers to present the different parts of Questions 1 and 2, perhaps having a couple of students discuss why Daphne's method works. Some students might give an algebraic argument based on the formula for circumference, perhaps even reviewing how they developed that formula using perimeters of polygons. Others might offer an intuitive geometric argument, perhaps using similarity, to explain why the circumference is proportional to the radius.

Questions 3 and 4

Questions 3 and 4 may require more discussion, because Daphne's plan doesn't work for area. Let students present the numerical results (Questions 3, 4a, and 4b), and then discuss why the plan doesn't work. Again, there may be a mixture of algebraic and geometric explanations.

Be sure to ask, What might Daphne have done to get the area? (given that she didn't know the area formula). Try to get students to articulate that she should have multiplied the area of her small circle by 50^2 rather than by 50. If necessary, ask how the actual area of the large circle compares to the area of the circle Daphne measured.

Algebra and Geometry

If students do not do so on their own, you should ask, How are Daphne's results related to the formulas for circumference and area? ($C = 2\pi r$ and $A = \pi r^2$).

Be sure students understand that the circle Daphne drew is the case $r = 1$, and to use this opportunity to introduce the term **unit circle**.

The unit circle has a circumference of 2π. According to Daphne's experience, multiplying its radius by a number r also multiplies the circumference by r. This gives a circumference of $2\pi r$, which matches the formula.

Similarly, the unit circle has an area of π. According to students' results, multiplying its radius by a number r multiplies the area by r^2, giving an area of πr^2, which again matches the formula.

Analogy with Polygons

Students should relate the conclusions reached about circles to comparable principles for similar polygons. (You may want to point out that all circles are similar—that is, they have the same shape.)

Ask, What analogous principles can you state for the perimeter and area of similar polygons? Students might state results like these.

- For any two similar polygons, the ratio of their perimeters is equal to the ratio of any two corresponding sides.
- For any two similar polygons, the ratio of their areas is equal to the square of the ratio of any two corresponding sides.

Students might use the reasoning about "blown-up" linear units and square units that was used in the discussion of *Squaring the Circle* to explain these facts.

Key Questions

What might Daphne have done to get the area?

How are Daphne's results related to the formulas for circumference and area?

What analogous principles can you state for the perimeter and area of similar polygons?

Defining Circles

Intent

Students come up with the equation of a circle with an arbitrary center.

Mathematics

In *Defining Circles*, students build on the examples to get the standard form of the equation of a circle.

Progression

The activity gently moves students toward discovery of the equation of a circle not necessarily centered at the origin. Question 1 asks students to describe which trees will be watered by a sprinkler placed at (6, 2) that waters everything within a radius of 5. Question 2 asks them whether that sprinkler will reach a particular point. Questions 3 and 4 generalize the students' work, first by considering a blade of grass at (x, y) to be watered by the given sprinkler, and then by placing the sprinkler at (a, b) and giving it a pattern with a radius of r.

As this activity is discussed, make sure that students understand what the equation of a circle looks like when the center has one or more negative coordinates.

Approximate Time

25 minutes for activity (at home or in class)
15 minutes for discussion

Classroom Organization

Individuals, followed by whole-class discussion

Discussing and Debriefing the Activity

Ask students to come to an agreement in their groups on the answers for Questions 1 and 2.

You don't need to get a complete list of lattice points that fit Question 1, but do challenge one or two of the less obvious points, insisting that the student justify the assertion that the given tree gets wet (presumably by using the distance formula or the Pythagorean theorem).

Question 2 will be helpful in forcing students to use the distance formula or the Pythagorean theorem rather than simply estimate from a sketch.

Question 3

For Question 3a, students might simply state that the point should be within 5 units of (6, 2). Press them to express this more explicitly as an algebraic condition. If necessary, suggest that they examine their work on Question 2 for ideas. This should help them see that the point (x, y) will get wet if and only if

$\sqrt{(x - 6)^2 + (y - 2)^2} \leq 5$ [or equivalently, if and only if $(x - 6)^2 + (y - 2)^2 \leq 25$].

This is another good occasion to review the meaning of the phrase "if and only if."

Once students have developed the inequality for Question 3a, it should be fairly easy to get the corresponding equation for Question 3b. They might write this as

$\sqrt{(x - 6)^2 + (y - 2)^2} = 5$ or as $(x - 6)^2 + (y - 2)^2 = 25$.

Ask students to give you a geometric ("sprinkler-free") statement of what this equation represents. How can you describe geometrically the points that fit this equation? Guide them to state explicitly that this is the equation of a circle with center (6, 2) and radius 5.

If needed, check that students know exactly what a circle is by asking for a geometric definition. They should be able to describe a circle as the set of points whose distance from a given point (the *center*) is a particular number (the *radius*). Ask, What is the center of the circle represented by this equation? (This was discussed following the activity *In, On, or Out?*)

The equation $\sqrt{(x - 6)^2 + (y - 2)^2} = 5$ is essentially a literal translation of this definition into algebraic form, using the distance formula to express the distance from (x, y) to (6, 2).

Issues About Signs

This is a good occasion to bring up the tricky issues about signs that often confuse students. For instance, ask, What is the center for the circle that is represented by the equation $\sqrt{(x + 6)^2 + (y - 2)^2} = 5$? It may help students to think of $x + 6$ as $x - (-6)$, so the center is (−6, 2).

Question 4

If students had trouble with Question 3, you may want to give them some time in groups to work on Question 4. If a hint is needed, ask what would need to change in the answer to Question 3 if the center or radius were changed. They should be able to come up with one or both of the equations

$(x - a)^2 + (y - b)^2 = r^2$

and

$\sqrt{(x - a)^2 + (y - b)^2} = r$

Tell students that mathematicians usually use the first of these two forms of the equation, mainly to avoid the square-root sign. Thus, $(x - a)^2 + (y - b)^2 = r^2$ is the *standard form* for the equation of a circle.

Ask students to take the earlier example, $\sqrt{(x - 6)^2 + (y - 2)^2} = 5$, and rewrite it in this standard form. Ask, How can you rewrite the earlier equation without the square-root sign? The key step is for them to realize that they need to square the radius (5) as well as remove the square-root sign.

Once they have written the equation as $(x - 6)^2 + (y - 2)^2 = 25$, review again what this equation represents. As needed, go over the fact that the numbers 6 and 2 come from the coordinates of the center and that 25 is the square of the radius.

Key Questions

How can you describe geometrically the points that fit this equation?

What is the center of the circle represented by this equation?

What is the center for the circle that is represented by the equation

$\sqrt{(x + 6)^2 + (y - 2)^2} = 5$?

How can you rewrite the earlier equation without the square-root sign?

Supplemental Activities

Right in the Center (reinforcement) contains problems similar to *Proving with Distance–Part I*, as well as a problem that asks students to find the equation of a circle having the center and radius found in the previous questions.

Hypotenuse Median (extension) does not deal with circles at all, but generalizes a principle concerning the median to the hypotenuse that is illustrated in the first part of *Right in the Center*, so it is a good follow-up to that activity.

The Standard Equation of the Circle

Intent

In this activity, students learn how to transform the equation of a circle into the standard form.

Mathematics

Students have just seen how to write the equation for a circle if they are given the center and the radius. Now they will look at how to find the center and radius if they are given an appropriate equation in a different form. This uses the process known as **completing the square**.

Progression

The textbook pages for this activity are intended as reference materials for the students to use after the whole-class discussion described below. This discussion begins with an example of an equation of a circle in standard form and asks students to expand and simplify the terms so that no parentheses remain. They are then given a similar example of an equation of a circle that is not in standard form and are asked to transform it into the standard form by completing the square. We recommend that the students be guided in this activity by whole-class discussion as they work.

Approximate Time

20 minutes

Classroom Organization

Individuals, guided by whole-class discussion

Doing the Activity

Have students work in groups and ask, How can you rewrite $(x - 6)^2 + (y - 2)^2 = 25$ without parentheses? They should simplify their answer if possible. Then have a student present the result, saving the steps involved for subsequent discussion. (If needed, review techniques for multiplying binomials as well as issues about signs.)

For example, the sequence might look like this.

$$(x - 6)^2 + (y - 2)^2 = 25$$
$$x^2 - 12x + 36 + y^2 - 4y + 4 = 25$$
$$x^2 - 12x + y^2 - 4y + 40 = 25$$
$$x^2 - 12x + y^2 - 4y + 15 = 0$$

Emphasize that the final equation is equivalent to the original, so its graph must be the same circle, with center (6, 2) and radius 5, even though that description is not apparent from the final equation.

Then ask students, Suppose you started with the final equation. What could you say about the graph? Help them to see that they could undo each of the steps to get back to the original, standard form, but that it might not be obvious what steps to take if they didn't have the original.

Transforming an Example
Give the class a specific equation such as $x^2 - 8x + y^2 + 10y + 4 = 0$. Point out that this resembles the final equation from the previous sequence, so it seems likely that its graph is also a circle.

Ask students to work backwards from this example and see if they can write it in standard form. (We recommend that you have students work on this initial example as a whole class.)

As a hint, you might ask, What expression could you square to give the terms x^2 and $-8x$? You might also suggest that they examine the previous sequence to see how each step followed from the preceding one.

If necessary, insist that students try squaring various expressions, such as $x + 3$ or $x - 5$, until they find one that gives a middle term $-8x$. Ask, What is the constant term in the expansion of $(x - 4)^2$? What equivalent equation includes the full expression $x^2 - 8x + 16$? Help them to articulate that the constant term in the expression being squared is half the coefficient of x. Review the phrase **completing the square** to describe the process of seeing $x^2 - 8x$ as part of the perfect square $(x - 4)^2$.

Once students see that they need the expression $(x - 4)^2$ to get the terms x^2 and $-8x$, have them identify the remaining term of the expansion of $(x - 4)^2$. Then ask how they can rewrite $x^2 - 8x + y^2 + 10y + 4 = 0$ as an equivalent equation that includes the full expression $x^2 - 8x + 16$.

Most students find it easiest to use an "add the same thing to both sides" approach in transforming the overall equation. For instance, they can add 16 to both sides of the original equation to get $x^2 - 8x + 16 + y^2 + 10y + 4 = 16$. A similar step for y gives $x^2 - 8x + 16 + y^2 + 10y + 25 + 4 = 16 + 25$.

Finally, have them combine constant terms (subtracting 4 from both sides) and write the equation using the perfect squares as $(x - 4)^2 + (y + 5)^2 = 37$

What Does This Equation Represent?
Conclude the discussion of this example by asking what this final equation represents. Two issues will probably need emphasis.
- **The signs of the coordinates:** Students should realize that the plus sign in the term $y + 5$ means that the y-coordinate of the center is -5.

- **Interpreting the constant term:** Be sure they grasp that the radius of the circle is $\sqrt{37}$, not 37.

Key Questions

How can you rewrite $(x - 6)^2 + (y - 2)^2 = 25$ without parentheses?

Suppose you started with the final equation. What could you say about the graph?

What expression could you square to give the terms x^2 and $-8x$?

What is the constant term in the expansion of $(x - 4)^2$? What equivalent equation includes the full expression $x^2 - 8x + 16$?

Supplemental Activities

Not Quite a Circle (extension) engages students in an open-ended exploration of variations on the equation of a circle that lead instead to an ellipse.

What's a Parabola?, Creating Parabolas, Coordinate Ellipses and Hyperbolas, Another View of Ellipses and Hyperbolas, Ellipses and Hyperbolas by Points and Algebra, Generalizing the Ellipse, and *Moving the Ellipse* (extensions) together form an intensive study of conic sections, primarily from a coordinate geometry perspective.

Completing the Square and Getting a Circle

Intent

Students practice transforming equations into the standard form for the equation of a circle in order to determine the center and radius.

Mathematics

This is continued work with transforming certain equations into the standard form of a circle. Students clarify the mechanics of completing the square and learn to deal with such difficulties as odd linear coefficients and equations having no solutions.

Progression

The activity presents students with four equations. They are asked to transform each into the standard form of the equation of a circle and to identify the center and radius of the circle. Question 3 introduces the difficulty posed by an odd linear coefficient, and Question 4 presents an equation that does not define a circle (where r^2 would be negative).

Approximate Time

30 minutes for activity (at home or in class)
10 minutes for discussion

Classroom Organization

Individuals, followed by whole-class discussion

Doing the Activity

If assigning this as homework, you may want to give students time to begin this in class, where you can provide some assistance with the process of completing the square if needed.

Discussing and Debriefing the Activity

Let students spend a few minutes comparing notes in their groups and then have a student present each problem.

The first two problems are similar to the example discussed in class with *The Standard Equation of the Circle*. You may want to anticipate the issue of dealing with odd coefficients (in Question 3) by asking, How did you know to use $(x - 4)^2$ to get the terms x^2 and $-8x$? As you did previously, help them to articulate that the constant term in the expression being squared is half the coefficient of x.

Question 4: A "Disappeared" Circle

On Question 4, students were probably able to complete the squares and transform the equation into $(x + 3)^2 + (y + 1)^2 = -3$. But they may have been confused about how to interpret this equation.

If necessary, ask, *Can you find any numbers for x and y that fit this equation?* Be sure to have students explain why this equation has no solutions. Emphasize that this also means that there are no points on the graph of the equation.

You may want to look at some variations on the equation in Question 4 to see how changes in the constant term affect the final result, perhaps starting from a similar equation in standard form that does represent a circle, such as $(x + 3)^2 + (y + 1)^2 = 6$. Include discussion of the special case $(x + 3)^2 + (y + 1)^2 = 0$, whose graph is the single point $(-3, -1)$. Students might describe this as a circle of radius 0.

Key Questions

How did you know to use $(x - 4)^2$ to get the terms x^2 and $-8x$?

Can you find any numbers for x and y that fit this equation?

Lines of Sight

Intent

Students assemble the final pieces of the puzzle in order to complete the unit problem.

Mathematics

At this stage, students have discovered how to measure the distance from a point to a line in order to determine the tree radius that will block a given line of sight in the orchard, and how to determine the time that will be required for the trees to grow to that radius. Now they will tackle the remaining question of determining which line of sight is the last to be blocked. They will then turn to solving the unit problem.

Formulas for the volume and surface area of cylinders are also introduced.

Progression

The Other Gap and *Lines of Sight for Radius Six* deal with the issue of identifying the last line of sight for the orchard. Students apply the principle from those activities in *Orchard Time for Radius Three*, where they also learn the distance between trees in the orchard. In *Hiding in the Orchard*, they finally put everything from the unit together to solve the unit problem.

Cylindrical Sodas prompts students to develop formulas for the volume and surface area of a cylinder, using their newfound formulas for the area and circumference of circles to extend their prior knowledge of surface area and volume of prisms to cylinders. *Big Earth, Little Earth* provides further reinforcement with the circumference formula for a circle, in the context of a problem with a surprising result.

The unit portfolio caps off the unit with reflection and summarization of lessons learned.

The Other Gap
Cylindrical Soda
Lines of Sight for Radius Six
Orchard Time for Radius Three
Hiding in the Orchard
Big Earth, Little Earth
Beginning Portfolios
Orchard Hideout Portfolio

The Other Gap

Intent

Students consider alternative lines of sight for the unit problem.

Mathematics

Students have now discovered how to find the tree radius that will block a particular line of sight in their orchard (*Cable Ready*) and how to calculate the time it will take for the trees to reach a given radius (*Orchard Growth Revisited*). This activity begins the exploration of the remaining task, identifying the last line of sight. At the same time, it gives students additional practice with finding the distance from a point to a line, although with a simplified situation in this case.

Progression

The Other Gap asks students to find how big the tree trunks must become in order to block the line of sight adjacent to the horizontal axis in a radius 3 orchard. The activity is introduced with a discussion of how, due to symmetry, there are really only two lines of sight that need be considered in this orchard. One of those lines of sight is identical to the path of the cable in *Cable Ready*, so this problem focuses on the other line of sight. The concluding discussion centers on recognizing which of these two lines of sight will be the last one to be blocked.

Approximate Time

5 to 10 minutes for introduction
20 minutes for activity
10 minutes for discussion

Classroom Organization

Individuals, followed by whole-class discussion

Materials

Large-grid poster paper
The Other Gap blackline master

Doing the Activity

The discussion here is intended to bring out two principles.

- An orchard may have several "families" of lines of sight.
- Among all lines of sight between two given trees, the optimal line is the one through the midpoint between the trees.

Students will use these two principles to find the hideout tree radius for a mini-orchard of radius 3.

You may wish to remind students that the "lines of sight" are really rays, but the geometric relationships do not change if we simply consider the lines containing these rays. This makes the language more familiar without altering the conclusions.

To determine if students see a connection between their work on the activity *Cable Ready* and the central unit problem, ask, What is the connection between Cable Ready and the central unit problem? If a hint is needed, you might suggest that they think of the cable as being like a line of sight.

One element of the connection is that both *Cable Ready* and the unit problem involve the distance from a point (a tree) to a line (the cable or a line of sight).

Remind students that early in the unit (in *More Mini-Orchards*), they considered the question of finding the hideout tree radius for a mini-orchard of radius 3. They probably did not reach a clear conclusion at that time, except that it was less than half the unit distance (because trunks of that size would actually trap Madie and Clyde).

Possible Lines of Sight
Have students sketch a mini-orchard of radius 3, and ask what they might consider for the "last" line of sight out of the orchard (from the origin).

They should see that by symmetry, there are two families of gaps between trees. (This was discussed following *More Mini-Orchards*, but probably needs reviewing.) As indicated in this activity, every gap between trees is like one in the first quadrant, and every gap in the first quadrant is like one of the two gaps shown in the next diagram. Thus, some lines of sight are like those that go through the gap between the trees at *A* and *B* (the darkly-shaded area), and others are like those that go through the gap between the trees at *B* and *C* (the lightly-shaded area). The diagram shows one line of sight in each of these gaps. (An enlarged copy of this diagram, without the lines of sight, is provided in *The Other Gap* blackline master.)

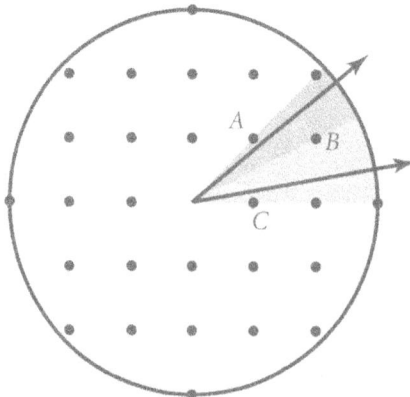

For an orchard of radius 3, every line of sight is like a line of sight in one of these two gaps.

Focus students' attention on the darkly-shaded gap in this diagram. Ask, Do you have to consider all lines of sight? What gaps between trees are like others? Bring out that the trees at A and B [that is, at (1, 1) and (2, 1)] are the key trees for this gap, because any line of sight that is blocked by the tree at (2, 2) will also be blocked by the tree at (1, 1). You can say that the trees at A and B "frame" the gap. Similarly, the trees at B and C frame the lightly-shaded gap.

Ask, What line of sight for the gap between (1, 1) and (2, 1) is the last to be blocked? Be sure students recognize that for such a line of sight to remain unblocked for as long as possible, it should be equidistant between these two trees. As needed, review the ideas from *Down the Garden Path*, in which students saw that such a line should go through the midpoint between the two trees. In other words, the best line of sight for this gap is the line that goes through the point $(1\frac{1}{2}, 1)$.

Ask, What do you know about this line? Help them to see that this is the same as the line along which the cable ran (in *Cable Ready*). You should have the conclusion from *Cable Ready* available, which says that this line of sight will become blocked when the tree trunks reach a radius of about 0.28 units.

The Other Gap
In this activity, students need to find out how big the tree trunks need to become to block the best line of sight for the other gap—the gap between the trees at (1, 0) and (2, 1). You can have students read the introduction to this activity as a summary of the preceding discussion.

Note: This problem can be analyzed in a somewhat simpler way than the *Cable Ready* problem because one of the trees under consideration here is on the horizontal axis. Details are given below.

Discussing and Debriefing the Activity

Have one or two students present their answers to the activity's question: How big must the tree trunks become to block this line of sight?

The following diagram shows the trees at (2, 1) and (1, 0), their midpoint D at $(1\frac{1}{2}, \frac{1}{2})$, the line of sight from the center through D, and the perpendicular segment from (1, 0) to the line of sight. Point E is the foot of this perpendicular, and the task is to find the length of $\overline{CE}$.

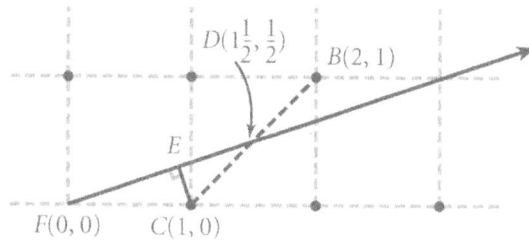

One approach uses the similarity between ΔFCE and ΔFGH, as shown here.

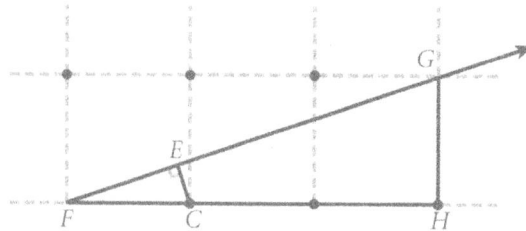

Comparing shorter legs and hypotenuses, we have

$$\frac{CE}{GH} = \frac{FC}{FG}$$

We have $FC = 1$ and $GH = 1$, and we can get $FG = \sqrt{10}$ by the Pythagorean theorem. Substituting gives $CE = \frac{1}{\sqrt{10}}$, which is approximately 0.32.

Another approach uses trigonometry, finding $\angle GFH$ from ΔGFH and then using that angle within similar ΔCFE.

The Hideout Tree Radius for the Orchard of Radius 3
Once students have presented as many methods as they have developed, pose this crucial question. Which line of sight is the last to be blocked?

Because $0.32 > 0.28$ (or, using exact values, because $\frac{1}{\sqrt{10}} > \frac{1}{\sqrt{13}}$), the line of sight just analyzed is clearly better. As already discussed, essentially there are only two choices available. The line of sight through $(1\frac{1}{2}, \frac{1}{2})$ is the last to be blocked, and the hideout tree radius (for the orchard of radius 3) is about 0.32 units.

Note: The use of the inequality $\frac{1}{\sqrt{10}} > \frac{1}{\sqrt{13}}$ is a good illustration of why "rationalizing denominators" often creates more confusion than clarity. In this case, it's easier to see that this inequality holds true than that the statement $\frac{\sqrt{10}}{10} > \frac{\sqrt{13}}{13}$ holds true. Situations like this, along with the increased role of technology for computation, are among the reasons why this topic is deemphasized in current curricula.

Key Questions

What is the connection between *Cable Ready* and the central unit problem?

Do you have to consider all lines of sight? What gaps between trees are like others?

What line of sight for the gap between (1, 1) and (2, 1) is the last to be blocked?

What do you know about this line?

Which line of sight is the last to be blocked?

Cylindrical Soda

Intent

Students develop formulas and principles for the volume and lateral surface area of a cylinder.

Mathematics

Students will use what they have learned about the circumference and area of a circle to find the volume and lateral surface area of a cylinder. They will also continue the theme from *Daphne's Dance Floor* by exploring how changes in a given dimension of a cylinder affect other measurements of the cylinder.

Progression

The first two questions of *Cylindrical Soda* ask students to find the volume and surface area of a given cylinder by applying principles from the formulas for volume and surface area of a prism. The final questions explore the effect on volume and surface area of scaling either the height or the radius of the cylinder.

Approximate Time

25 minutes for activity (at home or in class)
10 minutes for discussion

Classroom Organization

Individuals, followed by whole-class discussion

Discussing and Debriefing the Activity

You can assign students from different groups to go over parts of the solution. As they give individual numerical results, use the patterns and reasoning to bring out these general principles.
- The volume of a cylinder with height h and base of radius r is $\pi r^2 h$.
- The lateral surface area of a cylinder with height h and base of radius r is $2\pi rh$.
- Multiplying the height of a cylinder by a given factor multiplies both the volume and the lateral surface area by that factor.
- Multiplying the radius of the base of a cylinder by a given factor multiplies the lateral surface area by that factor but multiplies the volume by the square of that factor.

Supplemental Activity

Knitting (extension) extends this examination of the effects of scaling radius to the volume of a sphere.

Lines of Sight for Radius Six

Intent

Students identify the last line of sight for the general case.

Mathematics

Lines of Sight for Radius Six requires students to identify the last line of sight for the orchard of radius 6. The discussion generalizes this example and applies the generalization to the orchard of radius 50.

Progression

This activity is intended as a brief, intuitive examination of possible lines of sight for the orchard of radius 6. It presents students with a drawing of the first quadrant of that orchard, which they can explore using spaghetti or string to represent a line of sight. The subsequent discussion confirms (but does not prove) their intuition and then applies this conclusion to the orchard of the unit problem.

Approximate Time

5 minutes for introduction
5 to 10 minutes for activity
15 minutes for discussion

Classroom Organization

Individuals, followed by whole-class discussion

Materials

Spaghetti or string to represent lines of sight—at least one piece per group
Lines of Sight for Radius 6 blackline master: Copies for students and a transparency

Doing the Activity

A copy of the diagram is provided in the *Lines of Sight for Radius 6* blackline master, and you may want to reproduce this so that students can draw in the lines of sight. Alternatively, they can simply use pieces of spaghetti or string with the diagram in the textbook.

Bring out that to solve the unit problem, students will need to find the last line of sight for the orchard of radius 50. Tell them that this activity should help them locate it. Have students spend about five minutes working on this, and then bring them together for a discussion.

Discussing and Debriefing the Activity

Let students share their conclusions. Presumably, they will all agree that the last line of sight appears to be the one that goes through the gap between (1, 0) and (5, 1) [or one of the equivalent gaps, such as that between (0, 1) and (1, 5)].

Ask, *Can you name a specific point that this line goes through, other than the origin?* They should see that the line should go through the midpoint of these two points, which is at $(3, \frac{1}{2})$.

Students should recognize that they can do the same type of analysis here that they did in *Cable Ready* and *The Other Gap*. Assure them that this will confirm their intuition that this line of sight is the last to be blocked for the orchard of radius 6.

The Biggest Gap in General
Ask students to try to generalize this conclusion. They might say something like, **The last line of sight is the one through the 'lowest' gap in the first quadrant.**

Tell them that their intuition is correct but that a general proof of that fact is very advanced and involves mathematical concepts that they have not yet studied. Therefore, they must accept your word that their intuition about the largest gap is correct.

The Orchard of Radius 50
Finally, have students identify the last line of sight for the orchard of radius 50. They should see that it goes through the gap between (1, 0) and (49, 1), and must go through the midpoint of the segment connecting these two points, which is (25, $\frac{1}{2}$), as shown in the schematic diagram below. (This diagram is not drawn to scale.)

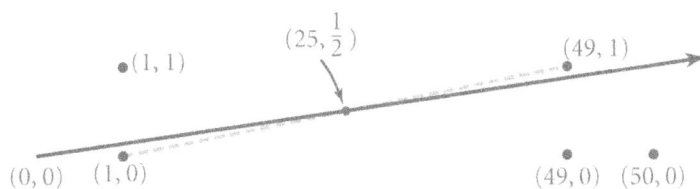

Key Questions

Can you name a specific point that this line goes through, other than the origin?

How would you generalize this conclusion?

What is the last line of sight for the orchard of radius 50?

Orchard Time for Radius Three

Intent

Students get another look at the issue of time for the orchard situation, before tackling the unit problem.

Mathematics

Students have done all of the mathematical components of this activity before, but this is the first time that they have put them all together. They must combine finding radius from circumference, finding area from radius, converting the measurement units, and utilizing the growth rate, in order to find the time required for trees to reach a given radius.

Progression

Students are given the current circumference of the trees and the growth rate, then are asked to find how long it will take for the last line of sight to be blocked in a radius 3 orchard. For the first time, they are also told that the unit distance in the orchard is 10 feet.

Approximate Time

25 to 30 minutes for activity
10 minutes for discussion

Classroom Organization

Individuals, followed by whole-class discussion

Discussing and Debriefing the Activity

You may want to have several volunteers each do part of the problem, as a way to see how well the class as a whole understands the situation.

The answer can be found using the following steps. As with the similar work in *Orchard Growth Revisited*, we recommend that you allow generous variation in results here, and that you not get distracted by the issue of significant digits.
*   The current radius of the trees is approximately 0.40 inches.
*   The current cross-sectional area of the trees is approximately 0.50 square inches.
*   The hideout tree radius is approximately 3.2 feet, which is approximately 38 inches.

- The trees need to reach a cross-sectional area of approximately 4500 square inches (so the initial cross-sectional area is negligible).

- It will take approximately $\dfrac{4500}{1.5} = 3000$ years for this growth to occur!

Clearly this is too long for Madie and Clyde. Fortunately, their orchard has a radius of 50 units, not 3.

Hiding in the Orchard

Intent

Students solve the unit problem.

Mathematics

Students will now put together the three major pieces of the unit problem: identifying the unit problem, finding the tree radius that first blocks that line of sight, and finding the time required for the trees to reach that size.

Progression

Hiding in the Orchard is the culmination of the unit. It essentially repeats *Orchard Time for Radius Three*, except that the orchard now has a radius of 50. Students need to find the distance from (1, 0) to the last line of sight for this orchard, then combine that figure with other results to solve the unit problem.

Approximate Time

40 minutes for activity
35 minutes for presentations (More time may be needed if you decide to have every group present.)

Classroom Organization

Individuals work on their write-ups, followed by groups preparing for presentations

Materials

Poster graph paper for groups that are presenting

Doing the Activity

Because this activity is to be included in students' portfolios, each student should do a complete write-up of the problem, including explanations of all calculations.

In addition, each group should prepare an explanation of its solution to be presented to the class. You may want to give each group a collective grade for its work on this activity.

Discussing and Debriefing the Activity

We suggest that you choose a group at random for the presentation and then choose an individual at random to act as presenter for the group. You may want to have several presentations of this final problem, looking for as many variations as possible in the method of solution.

It turns out that the distance from (1, 0) to the last line of sight is $\dfrac{1}{\sqrt{2501}}$ units (approximately 0.02 units). This means that the hideout tree radius is about 0.2 feet, which is 2.4 inches. Students will probably be surprised that so small a radius will block all the lines of sight.

As before, students should find that the trees had an initial radius of about 0.40 inches and an initial cross-sectional area of about 0.50 square inches.

The trees need to grow to a cross-sectional area of about $\pi(2.4)^2$ ($\approx$ 18.1) square inches, so they have about 17.6 square inches of cross-sectional area to go. Since the cross-sectional area increases by about 1.5 square inches each year, it should take about $\dfrac{17.6}{1.5}$ years, or about 11 years 9 months, to create a hideout. This is a long time for a small change in radius, but not nearly so bad as the 3000 years for an orchard of radius 3.

Big Earth, Little Earth

Intent

Students apply the circumference formula for a circle.

Mathematics

Big Earth, Little Earth presents a problem about circumference whose conclusion is rather surprising to most people.

Progression

The activity proposes stretching string around the circumference of the earth and around a six-inch radius globe. The string is then lengthened in each case by one foot. The students are asked to find the length of the original strings and then to predict which radius would have to grow more to accommodate the larger circumference. Finally, they are instructed to justify their answer. The subsequent discussion uses algebra to bring out why the result is the same for both spheres.

Approximate Time

25 minutes for activity (at home or in class)
15 minutes for discussion

Classroom Organization

Individuals, followed by whole-class discussion

Discussing and Debriefing the Activity

Students (and some teachers) are usually surprised by the fact that both radii in this problem would need to grow by the same amount ($\frac{1}{2\pi}$ feet, which is about 2 inches). Of course, the 2 inches will seem significant when added to a 6-inch radius, and it will seem inconsequential when added on to 4,000 miles. This discrepancy in the *significance* of the change may explain why our intuition leads us astray.

You can have students come to agreement in their groups on the solution. Then have them write an expression for their work without actually doing the arithmetic. This will help them see the distributive law at work, which is part of what this problem is about.

Here we present the analysis as it might be done without actually carrying out the arithmetic.

For the earth, the original radius is 4000 · 5280 feet, so the circumference is 2π · 4000 · 5280 feet. This means that the new length for the string is [(2π · 4000 · 5280) + 1] feet, so the new radius is

$$\frac{(2\pi \cdot 4000 \cdot 5280) + 1}{2\pi}$$

which is equal to [4000 · 5280 + $\frac{1}{2\pi}$] feet.

For the globe, the original radius is 0.5 feet and the circumference is 2π · 0.5 feet. This means that the new length for the string is [(2π · 0.5) + 1] feet, so the new radius is

$$\frac{(2\pi \cdot 0.5) + 1}{2\pi}$$

which is equal to [0.5 + $\frac{1}{2\pi}$] feet.

In each case, the increase over the original radius is $\frac{1}{2\pi}$ feet.

Ask one or two students to report on their group's work. If it doesn't come up in the presentations, you might have students work through what would happen using r as the original radius, so they see that the result can be generalized.

Beginning Portfolios

Intent

This is the first step for students in compiling their portfolios for this unit.

Mathematics

In *Beginning Portfolios,* students summarize what they have learned about coordinate geometry and about pi.

Progression

The write-ups for this activity will be required as part of the unit portfolio, described more fully in *Orchard Hideout Portfolio.*

Approximate Time

30 minutes for activity (at home or in class)
10 to 15 minutes for discussion

Classroom Organization

Individuals, followed by several student presentations

Discussing and Debriefing the Activity

Have several volunteers share the work they did on this activity. If students seem to need prompting, you might ask about the midpoint formula or the equation of the circle.

Orchard Hideout Portfolio

Intent

Students complete the unit portfolio.

Mathematics

The portfolio gives students an opportunity to reflect upon and summarize the various mathematical principles that they have studied in this unit. It can also serve as a nice reference for this material in their later studies.

Progression

Students will have done part of the selection process in *Beginning Portfolios*, so their main task in this assignment is to write their cover letters.

Approximate Time

45 minutes to an hour for portfolio completion (at home or in class)
10 minutes for discussion

Classroom Organization

Individuals

Doing the Activity

Be sure that students bring in their portfolios with the cover letters as the first item. They should also bring to class any other work that they think will be of help on the unit assessments. The remainder of their work can be kept at home.

Discussing and Debriefing the Activity

Let volunteers share their portfolio cover letters as a way to start a discussion to summarize the unit.

There are many ideas in the unit, from both coordinate and synthetic geometry. You can use the concepts and skills list in the unit overview for reference in this discussion.

Blackline Masters

More Mini-Orchards: Radius Two

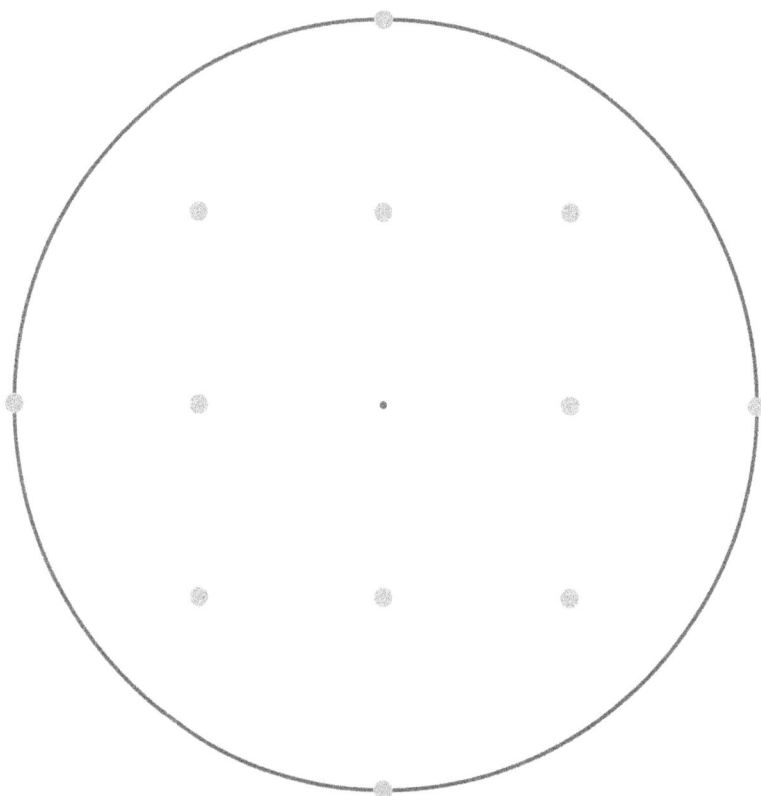

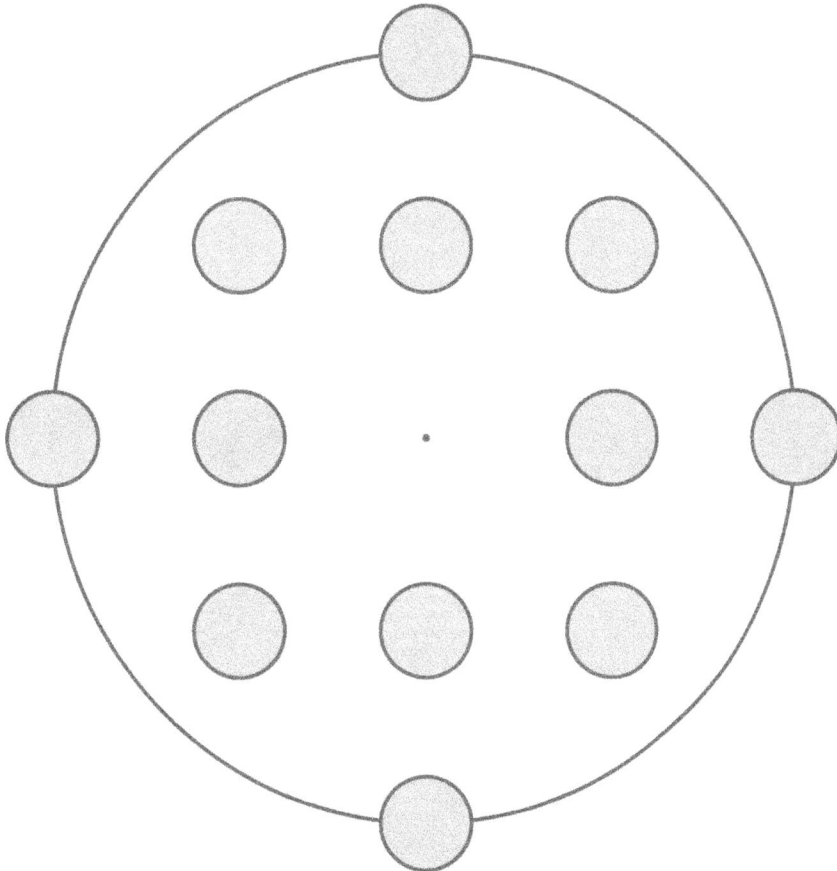

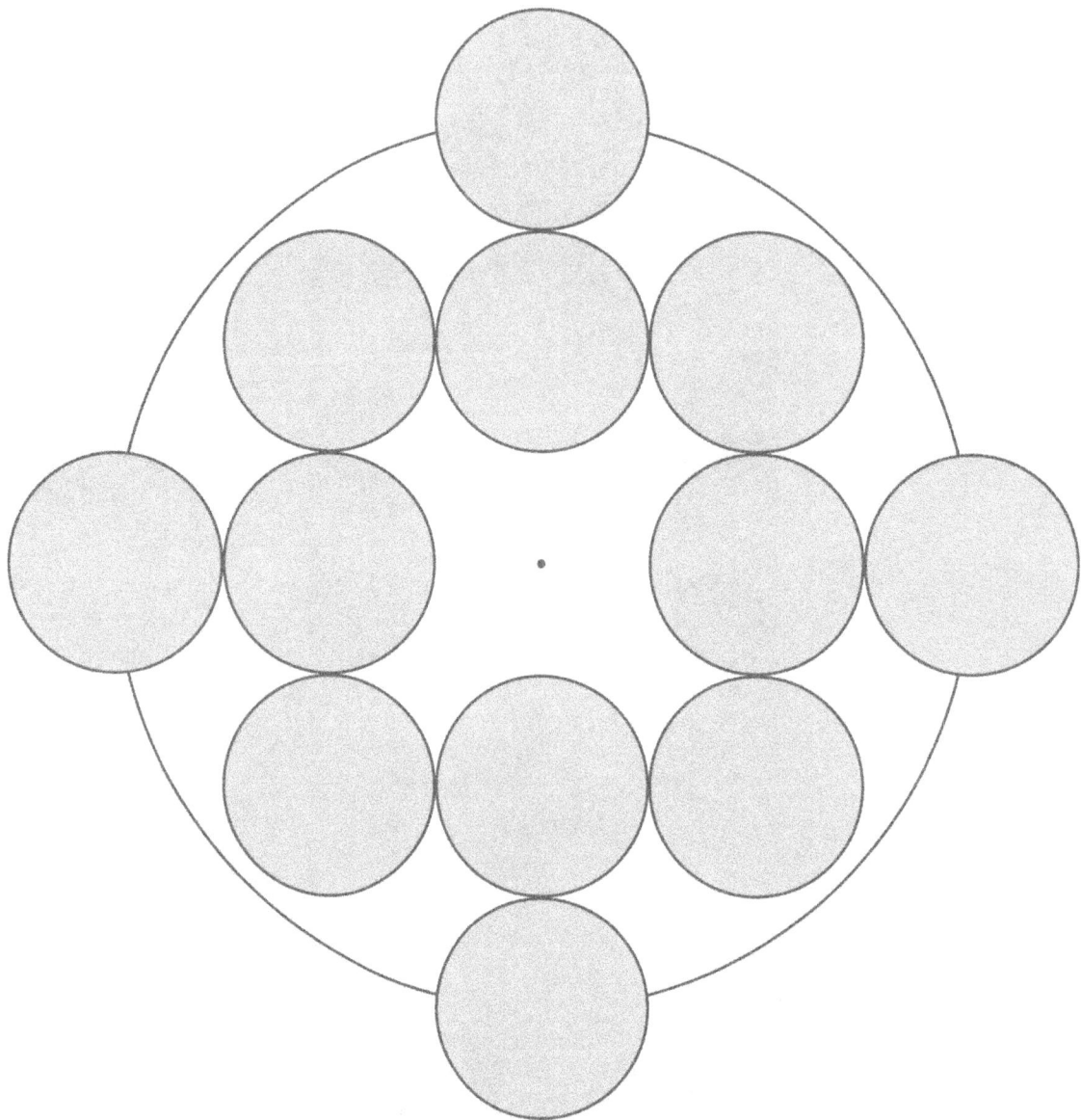

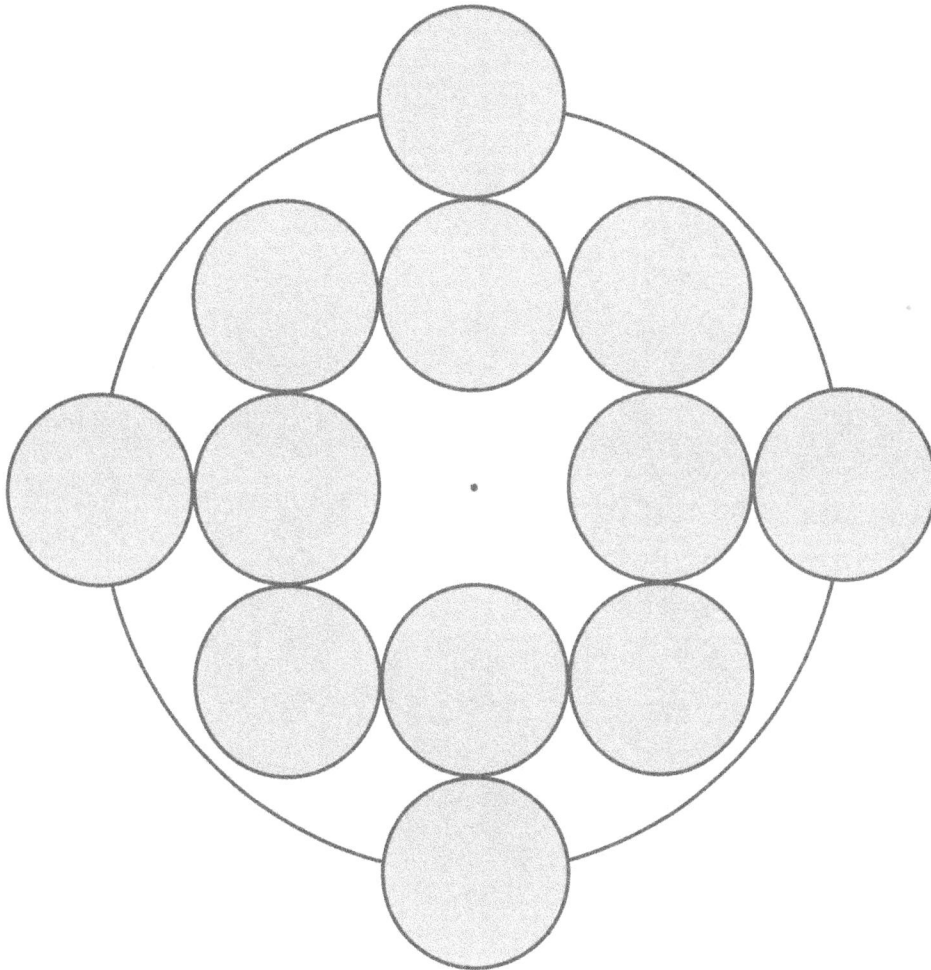

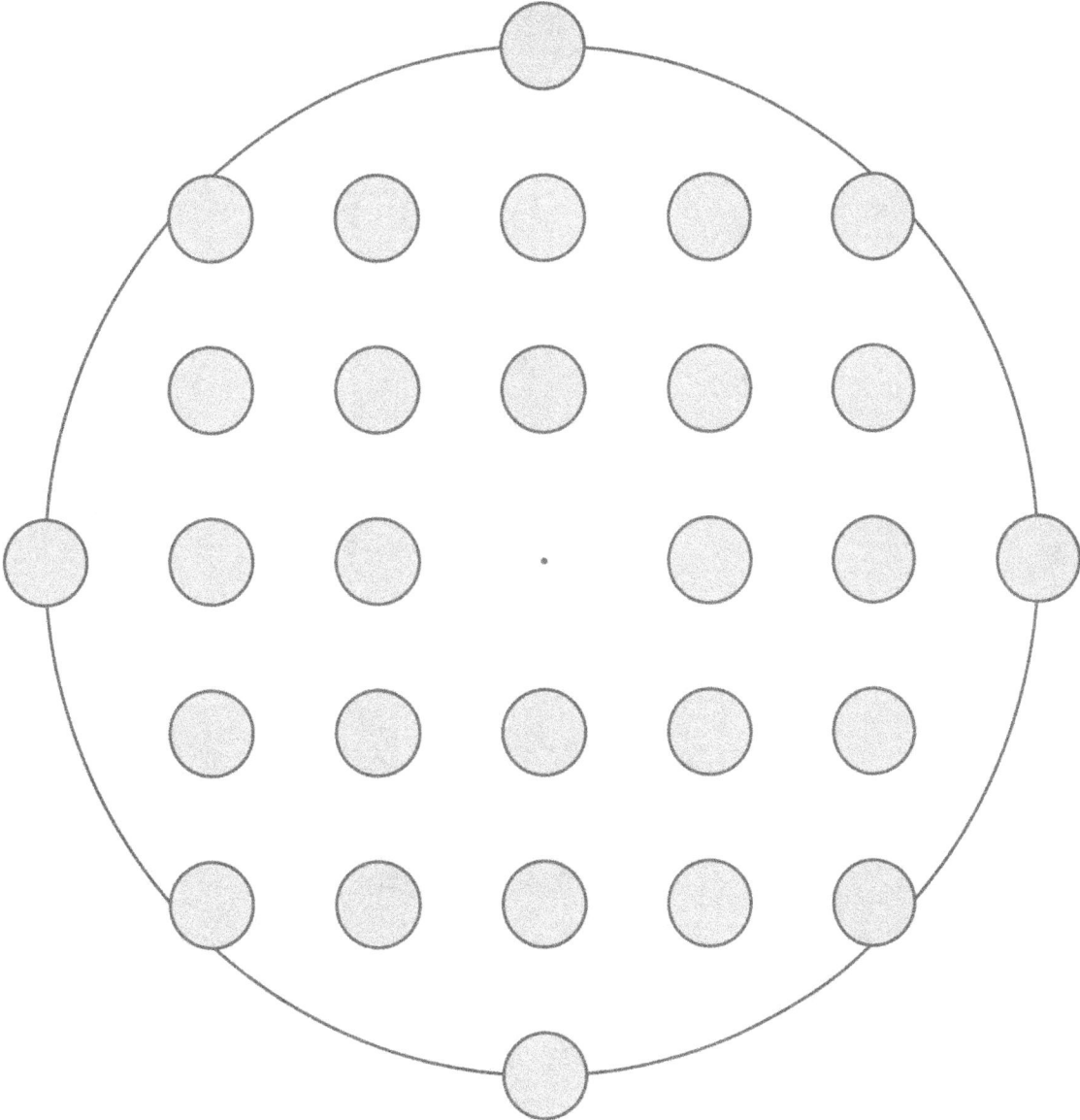

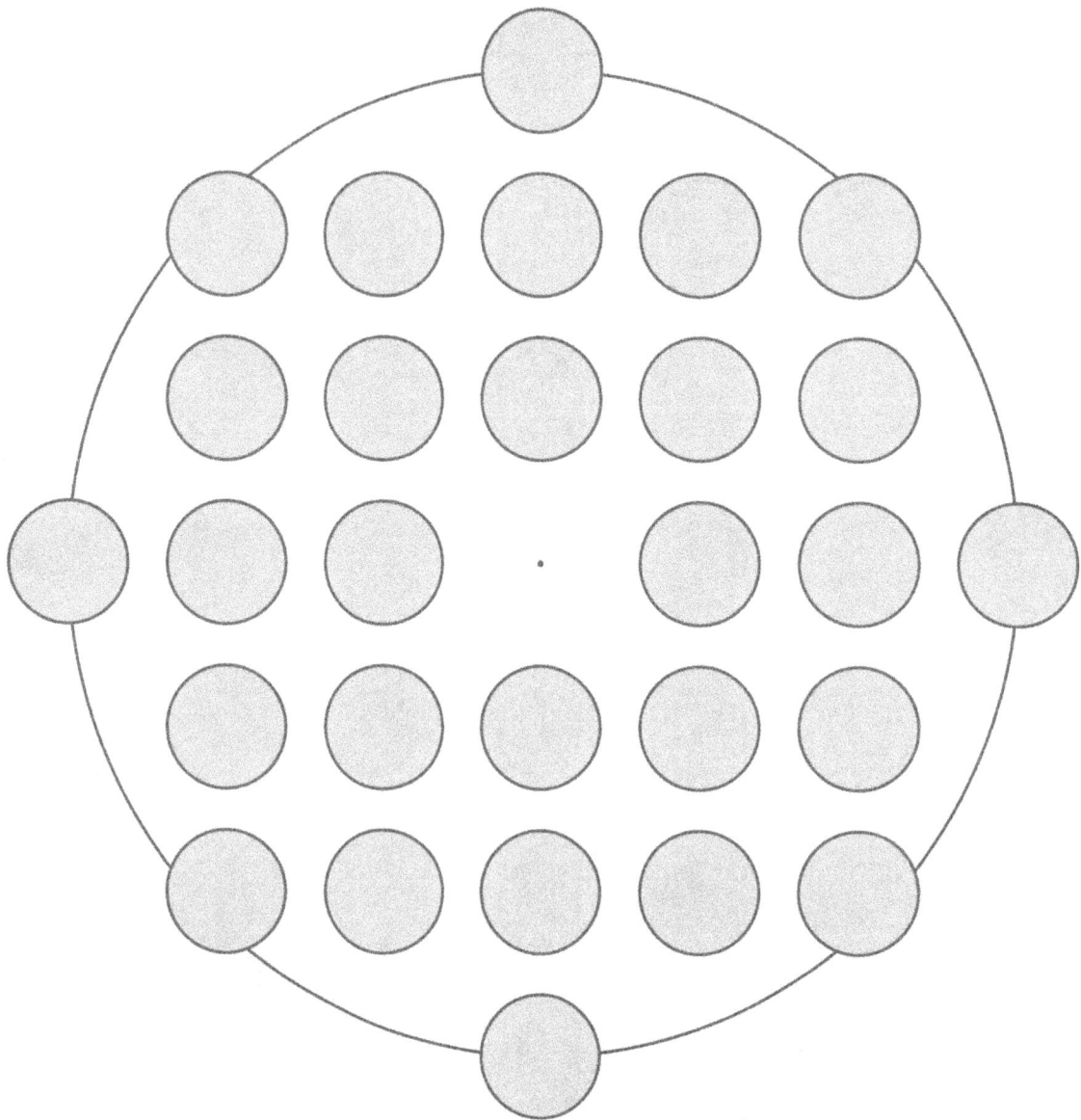

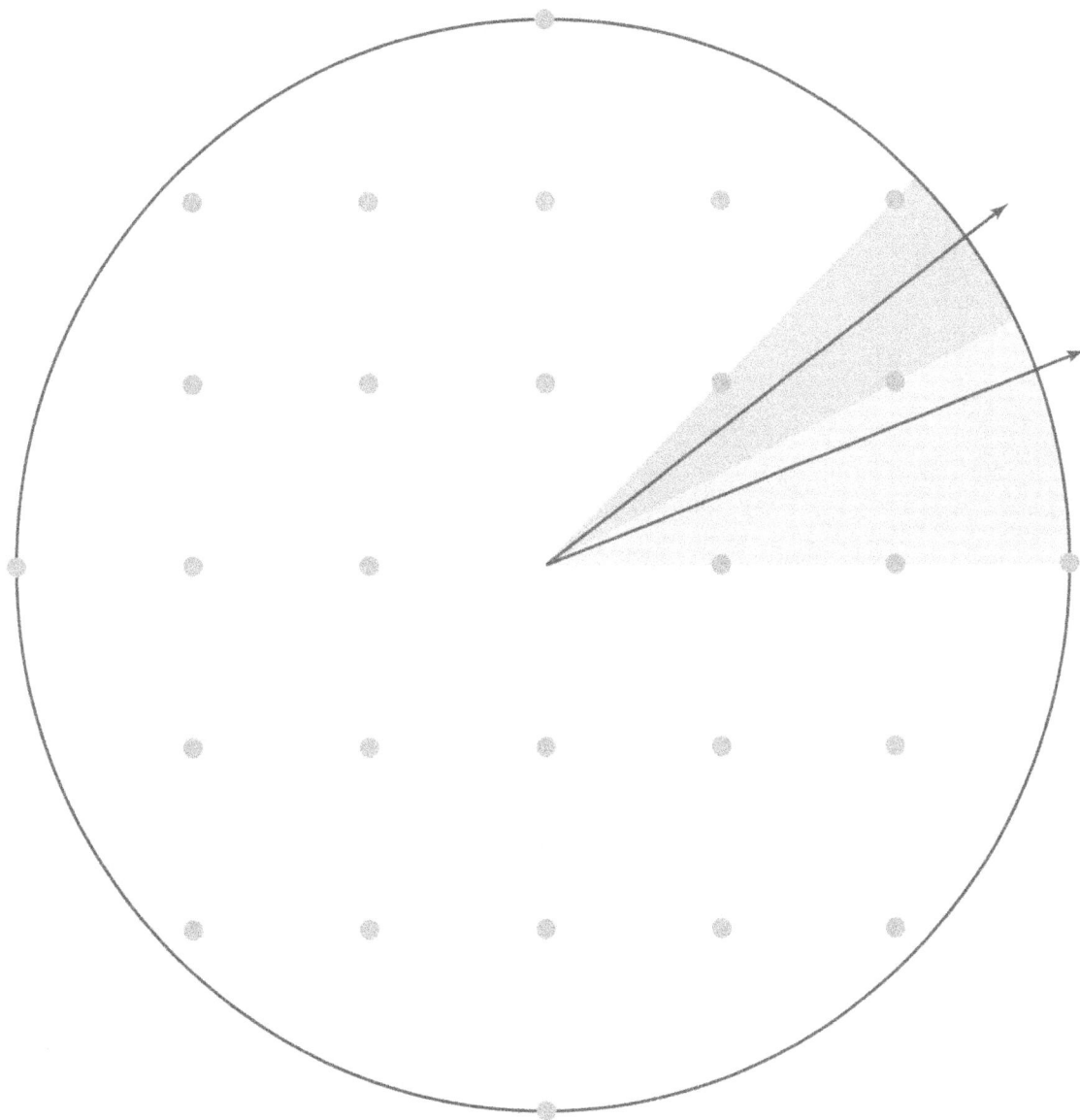

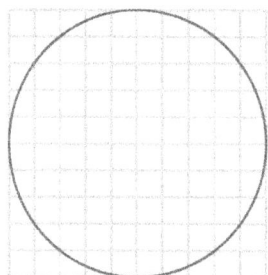

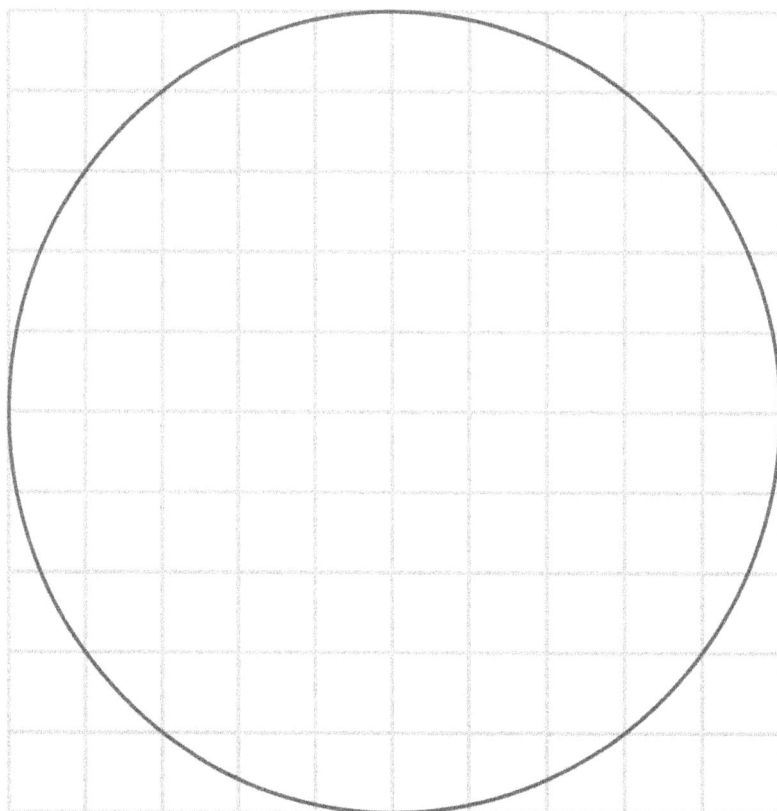

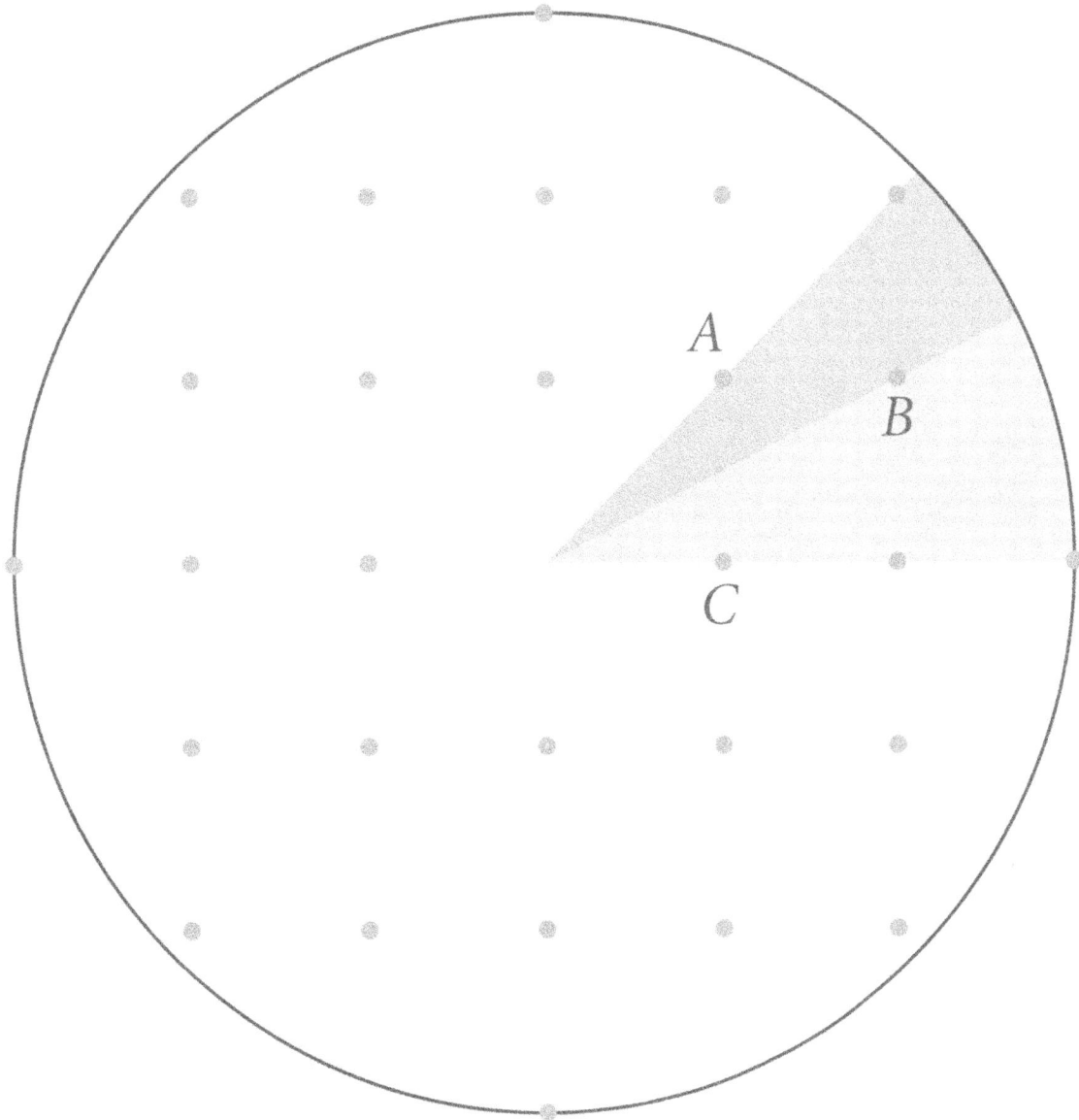

¼-Inch Graph Paper

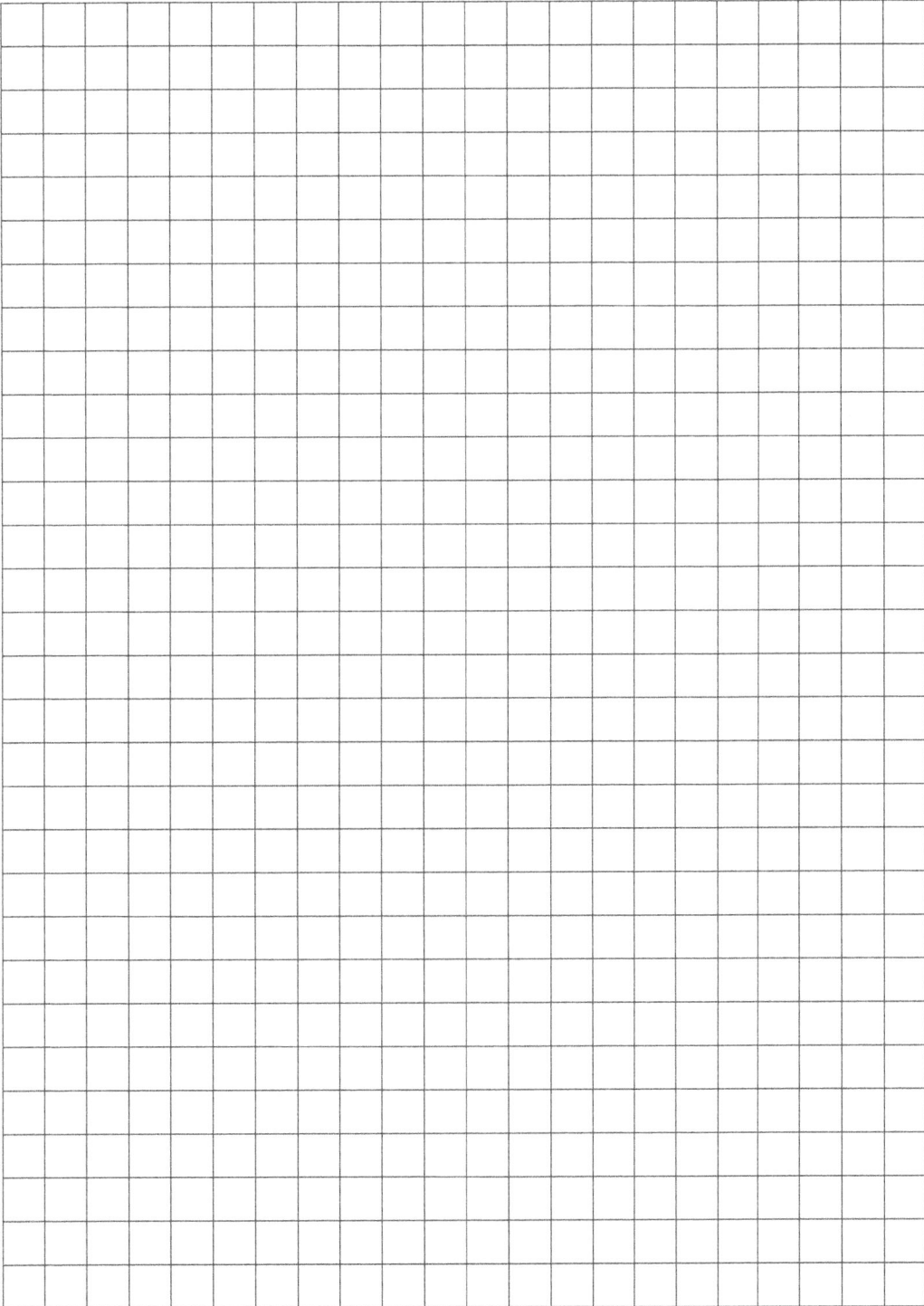

© 2011 Interactive Mathematics Program

1-Centimeter Graph Paper

1-Inch Graph Paper

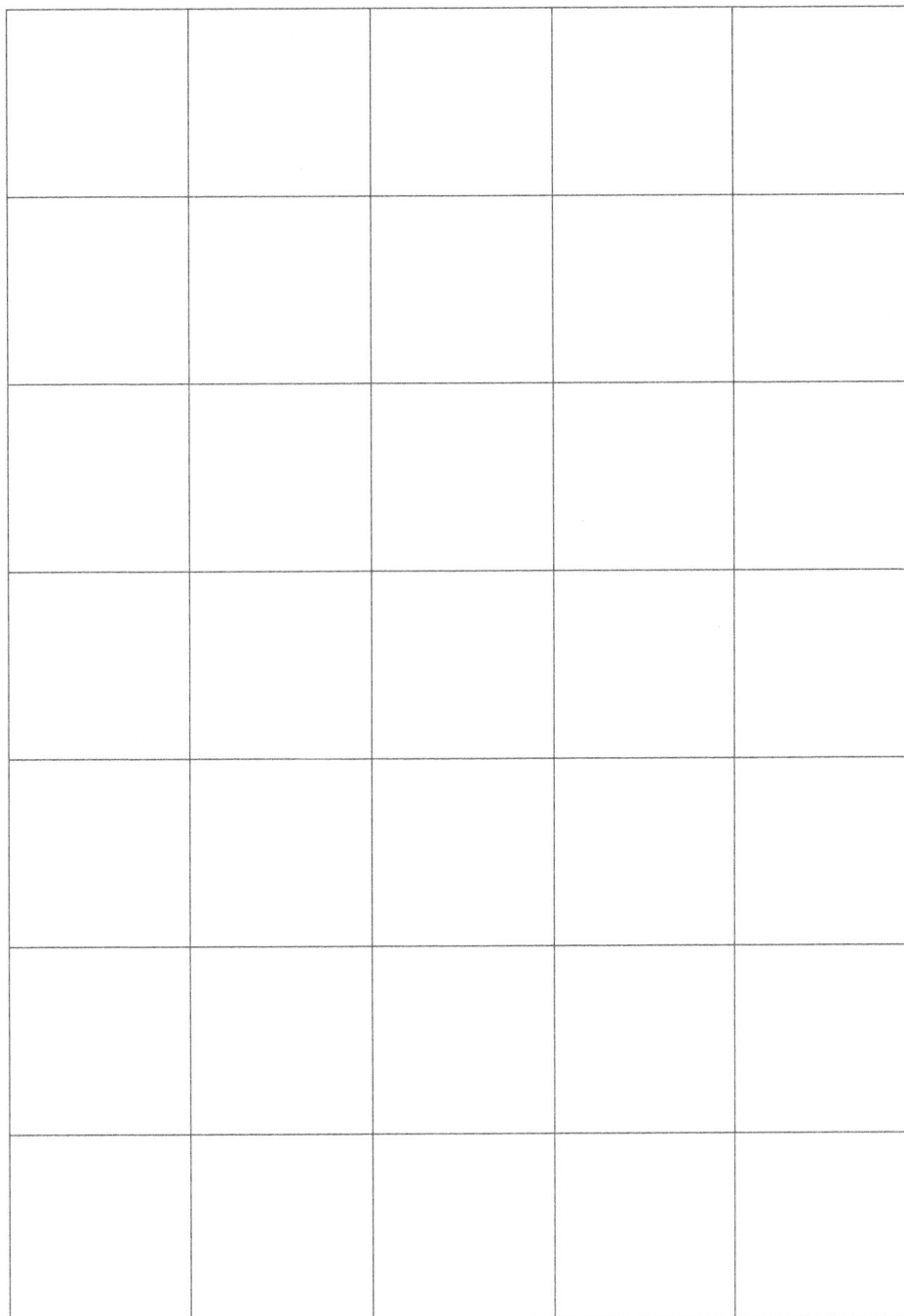

Assessments

In-Class Assessment

Madie and Clyde have decided that they would like to listen to music while they work in the orchard.

Clyde goes to the local electronics store and picks up a transmitter that will send an electronic signal through the air. The signal will reach up to 400 feet in every direction.

Clyde also buys two sets of headphones that will pick up the signal if they are close enough to the transmitter.

One day, Madie is working on the tree at (−26, −35) and Clyde is working on the tree at (46, −5). (The unit distance, such as the distance from (0, 0) to (1, 0), is 10 feet.)

1. Madie and Clyde want to place the transmitter so that the sound waves will reach them both with equal strength.
   a. Is this possible? Explain your reasoning.
   b. If so, where should the transmitter be located to be as close to them as possible? How far will the transmitter be from them?

2. Suppose that the transmitter is placed at the center of the orchard. Write an equation for the boundary of the broadcast region.

3. Suppose Madie takes the day off, and Clyde places the transmitter where he is working that day, at (7, !3). He decides to take a break and stroll around the orchard, and wants to know where he can go and still be able to hear the music. What is the equation of the boundary of the region that the sound waves will reach?

1. Hot Tub

Madie and Clyde have put a cylindrical hot tub in the center of their orchard. The radius of the circular base is 5 feet, and the tub is 3.5 feet deep.

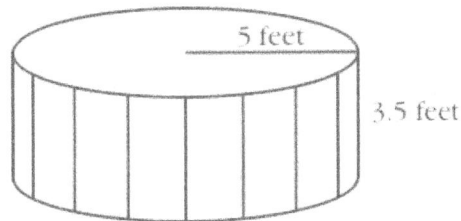

a. Clyde has made a canvas cover for the hot tub to keep the leaves out. The cover just fits across the top. How many square feet of material did he need for the cover? (Assume that no material was wasted.)

b. When the hot tub is completely filled, how many cubic feet of water will it hold?

c. Madie has decided to apply a sealant to protect the outside vertical surface of the hot tub against weathering. A quart of sealant will cover 100 square feet of surface. How many quarts will she use?

2. Orchard Archery

Madie and Clyde have taken up the sport of archery. They have uprooted the tree located at (–4, –15) and are using that location as the position from which they will shoot their arrows.

a. They plan to hang a target on another tree. They want the target to be located 5 units from the shooting point. (Assume that the tree trunks are still very narrow, so Madie and Clyde can treat the target as if it were at the center of the tree and other trees won't get in the way.)

   Give the coordinates of one possible tree where they can place the target. Keep in mind that they cannot shoot arrows through the remaining trees!

b. After some practice, Madie and Clyde decide that the target is too close for them, and they put up a new target at $(-13, -20)$. (They will still shoot from $(-4, -15)$.) But when their daughter Robin shoots, she gets to stand closer to the target than they do. As a matter of fact, she gets to stand halfway between their shooting point and the new target.

   What are the coordinates of the point from which Robin gets to shoot?

I. Once Upon a Time . . .

Imagine that you are Madie or Clyde. You've grown old and are telling your grandchildren the story of the orchard hideout. You've described the arrangement of trees in the original orchard (with a radius of 50 units) and told them the basic facts that you knew at the start.

* The circumference of the newly planted trees
* The fixed amount by which the cross-sectional area of the trees grew each year
* The distance between the centers of adjacent trees

Of course, the grandchildren have heard the story before, and they remember that it took about 11 years, 9 months for the center to become a true orchard hideout. What you want to do is impress them with how well you and your partner analyzed the problem back then.

Write a description of how the analysis worked. Don't get bogged down in the specific numbers, because you don't have pencil and paper handy, and the youngsters are more interested in the big ideas anyway!

II. Road Building

The highway department is planning a road that will go through the town of Coldwater. The town of Hot Springs is 13 miles due south of Coldwater, and the town of Warm Rock is 18 miles due east of Hot Springs.

1. Sketch a diagram showing the relationship between the three towns. (Treat each town as a single point.)

The mayors of Hot Springs and Warm Rock both want this new road through Coldwater to go straight through their towns as well. Unfortunately, the highway department can afford to build only one road.

The road must go through Coldwater and must be straight, so a compromise route is needed. The mayors of Hot Springs and Warm Rock agree to support the project if this condition is met:
The distance from Hot Springs to the new road must be the same as the distance from Warm Rock to this road.

They also insist that the road should not be parallel to the route from Hot Springs to Warm Rock.

2. a. Add a dotted line to your sketch from Question 1 to show where the new road must go, and explain your reasoning.

   b. Find the distance from Hot Springs to the road, to the nearest tenth of a mile, and explain your work.

III. Equation Time

Solve this system of equations, and explain your work.

$$7r + 6s = 6$$
$$5r - 4s = 25$$

IV. The Third Dimension

This system of linear constraints in three variables defines a feasible region.

I	$2x + y + z \leq 20$
II	$3x + 4z \leq 13$
III	$y + 2z \leq 9$
IV	$x \leq 9$
V	$y \leq 6$
VI	$x \geq 0$
VII	$y \geq 0$
VIII	$z \geq 0$

Give *a general outline* of how to find the point in this feasible region where the function $2x + y + z$ has its maximum, and explain the geometric reasoning behind your method.

V. Solving with Matrices

Consider this system of linear equations.

$$3a + 2b - c + d = 1$$
$$2a - b + 4c + 2d = -2$$
$$-4a + 3c - 3d = -6$$
$$a + b + c + d = 3$$

1. Write a matrix equation that is equivalent to this system.

2. Solve the system using matrices on a graphing calculator, and show your solution.

3. Discuss the relationship between the matrices and the equations and the properties of matrices that allow you to use them to solve systems of linear equations.

I.   Spilt Milk

You've automated your dairy farm so that all the cows are milked by milking machines, and the milk all flows into one giant cone-shaped container. At the start of milking time, the container is empty, and as the milk flows in, the level in the container rises. Milking starts at 5:00 a.m. and continues through the day. (The cows are not all milked at the same time.)

After studying your cows and using some geometry, you've figured out that at t minutes after 5:00 a.m., the milk in the container will have risen to a level of $\sqrt[3]{2000t}$ centimeters.

   1. During the hour from 7:00 a.m. until 8:00 a.m., what is the *average* rise per minute in the height of the milk? (Give your answer to the nearest 0.001 cm/min.)

   2. At what rate is the milk level rising at 8:00 a.m.? (Again, give your answer to the nearest 0.001 cm/min.)

   3. At what time of day will the milk level reach 100 centimeters?

II.   Darts

Consider a square dartboard with a circle inscribed in the square, as shown here.

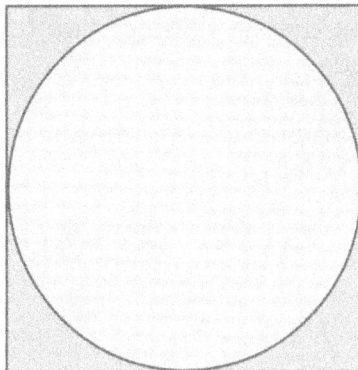

Suppose that according to the rules, if your dart lands inside the circle, you win, and if the dart lands outside the circle, you lose. Assume that you always hit the dartboard and that each point of the square is equally likely to be hit.

1. If you throw one dart, what is your probability of winning? Explain your answer, giving the probability to the nearest hundredth.

2. Suppose you throw seven darts. What is the probability that you will win at least four times? Explain what method you use to find the answer and why the method works. Again, give the probability to the nearest hundredth.

III. Ferris Wheel Fence, Revisited

It's time to look back at the problem of the fence around the amusement park, from *High Dive*.

As you may recall, Al and Betty are riding on a Ferris wheel. This Ferris wheel has a radius of 30 feet, and its center is 35 feet above ground level. There is a 25-foot-high fence around the amusement park, but once you get above the fence, there is a wonderful view.

What percentage of the time are Al and Betty above the level of the fence?

IV. Opposite Angles

You have learned these formulas involving trigonometric functions:

$$\cos(-\theta) = \cos\theta$$
$$\sin(-\theta) = -\sin\theta$$

Explain each of these formulas in several ways:

- In terms of the Ferris wheel
- In terms of the graphs of the sine and cosine functions
- Using numerical examples

You can use these graphs of sine and cosine in your explanation:

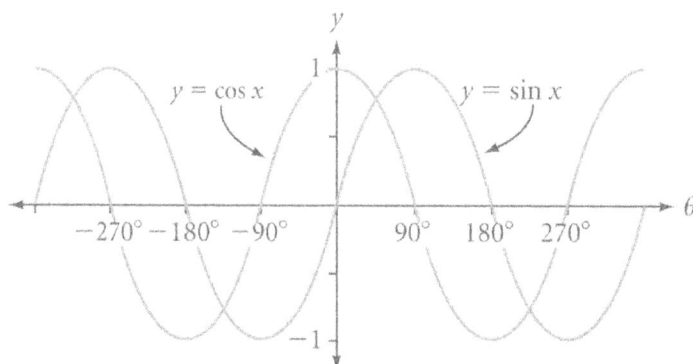

You can also use this diagram to represent a Ferris wheel:

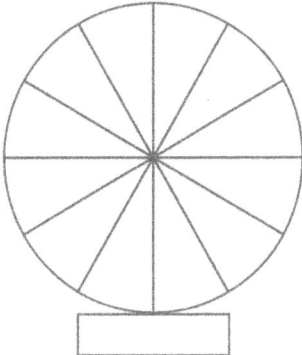

Orchard Hideout Calculator Guide for the TI-83/84 Family of Calculators

Orchard Hideout focuses mainly on geometric and algebraic concepts and does not require any significant new uses of the graphing calculator. However, the unit provides several new applications of familiar calculator features, as well as opportunities to explore new features. Students may wish to use the calculator's programming and graphing abilities while solving some problems, and the visual nature of the unit's geometry may inspire use of the calculator's drawing features as well.

Students will use the calculator daily for varied computational purposes, including use of the sine, cosine, and tangent functions. For a review of the calculator's basic features, see *Calculator Basics* in the Year 3 general resources.

Orchard Hideout: This activity encourages students to develop models of what the orchard might look like. Students may wish to use the drawing features of the graphing calculator to create a smaller version of the orchard. The Calculator Note "Drawing an Orchard" gives instructions for creating a simple model on the graphing calculator. You may find such drawing useful for class discussion about this activity, or later in the unit. But hold off on giving this Calculator Note to students until later in the unit, or give it only to students who request it.

Similar work with the drawing features will allow students to investigate trees of different sizes in later activities, such as *More Mini-Orchards*. When students do begin to draw orchards on the graphing calculator, give them also the Calculator Note "Storing and Recalling a Picture."

More Mini-Orchards: In this activity, students investigate what tree radii will create a hideout in smaller, simpler orchards than the one in the central unit problem. If students are interested in the drawing features of the calculator, encourage them to draw a model of a small orchard.

If you have students who are interested or skilled in programming, ask them to write a program that will generate orchards of various sizes. This would be a challenging project for an interested student, and the results might prove creative.

The Calculator Note "Programming an Orchard Hideout" gives specific instructions for a program, ORCHARDS, to draw various orchard and tree sizes. It is provided primarily for your convenience, rather than as instructions for students. You may find using the program to be a useful addition to class discussions. However, be aware that even orchards of a modest size will take quite some time to draw. You can download ORCHARDS from the Resources in the Orchard section of the *IMP Online Teacher's Guide*. *Calculator Basics* gives general instructions on how to transfer a program from computer to calculator.

Review the Calculator Note "Programming the Calculator," in *Calculator Basics*, for general instructions on how to program a TI graphing calculator.

***In, On, or Out?*:** In this activity, students develop an equation or equations to describe a circle, for example, $x^2 + y^2 = 10^2$. Students may wish to graph this on the graphing calculator. Doing so will offer an opportunity to revisit the definition of a function and to create the need for symbolic manipulation in order to enter this equation in the Y= editor.

Students will quickly discover that graphing **Y₁=√(10²–X²)** causes two problems. The first problem is that you see only half (the top half) of the circle. The equation $x^2 + y^2 = 10^2$ does not represent a function, but the equation $y = \sqrt{10^2 - x^2}$ does. Students may recognize that graphing **Y₂=–√(10²–X²)** will complete the circle. (Note that a similar topic, the standard equation of a circle, comes up in the activity *Completing the Square and Getting a Circle*.)

The second problem is that the circle may not look like a circle. The calculator displays a viewing window as set by the user in the **WINDOW** menu. Unless these window values are specifically chosen to create *x*-distances that are visually equal to *y*-distances, the graph appears stretched. You can easily fix this by pressing ZOOM and selecting **5:ZSquare**.

***The Distance Formula*:** When students discuss Question 4, which asks whether the distance formula holds true for points with negative coordinates, expect some to ask, "Why does my calculator show -3^2 as equal to -9?" The order of operations that the calculator follows will first square the value 3 and then apply the negative sign to the resulting square.

Encourage students to review what squaring means, and remind them that entering **(–3)²** gives the square of –3.

***Hexagoning the Circle*:** Students may use trigonometric ratios to help solve for the area and circumference of a hexagon. But they may question the reasonableness of some of their results. Remember, a potential trouble spot is that the mode may be set to radians rather than degrees. To set the calculator to degrees, press MODE and highlight **Degree**. Then press 2ND [QUIT] to return to the home screen.

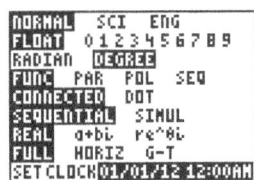

```
NORMAL  SCI  ENG
FLOAT  0123456789
RADIAN  DEGREE
FUNC  PAR  POL  SEQ
CONNECTED  DOT
SEQUENTIAL  SIMUL
REAL  a+bi  re^θi
FULL  HORIZ  G-T
SET CLOCK 01/01/12 12:00AM
```

***Polygoning the Circle*:** As noted in the *Teacher's Guide,* point out that pressing 2ND [π] will enter the calculators' approximation of π, which will lend precision to students' calculations.

***Marching Strip*:** In doing this POW, if students discover the associated greatest common divisor (GCD) idea, they may ask if their calculator has that feature. To find this command, press MATH, move to the right to highlight **NUM**, and scroll down to select **9:gcd(**. To complete the command, enter the two values for which you wish to find the GCD. For example, to find the GCD of 63 and 90, enter **gcd(63,90)** and press ⌷ENTER⌷. The result, 9, will be displayed.

Orchard Growth Revisited: The *Teacher's Guide* suggests reminding students to take advantage of the last-answer memory feature to use the calculator's multi-digit values from one calculation to the next. Although the focus of this activity is on geometry as opposed to the complexities of significant digits, students may find the Calculator Note "Using ⌷2ND⌷ [ANS] and the Last Entry" helpful.

In this activity, students may again have trouble ensuring that order of operations is followed when they use a calculator. They may mistakenly calculate 2.5/2π as equal to (approximately) 3.93 because they neglected to enclose 2π in parentheses. Enter **2.5/(2π)** to correctly determine the approximate decimal value of this fraction.

Supplemental Problems: The majority of the supplemental problems explore geometric ideas, and they are not dependent on graphing calculators. However, if students use calculators to calculate trigonometric ratios, remind them to check that their calculators are in degree mode.

In some of the later supplemental problems that deal with parabolas, ellipses, and hyperbolas, students may wish to use a graphing calculator to check their equations. Recall that they will need to solve for *y* to enter the equations in the Y= editor, and that if they take a square root, they'll need to enter both the positive and negative branches of the result separately. A quick shortcut to this is to enter the positive branch in Y_1, then enter $Y_2 = -Y_1$. To enter Y_1, press ⌷VARS⌷, scroll right to **Y-VARS**, press ⌷ENTER⌷ to select **1:Function...**, and press ⌷ENTER⌷ to select **1:Y_1**.

Drawing an Orchard

You can construct a simple drawing to represent an orchard using the **DRAW** and **GRAPH** features of the graphing calculator. You can also make adjustments to this "starter orchard" to represent orchards of different sizes. Keep in mind that there is a limit to the exactness of this model: it is restricted by the location of the pixels on the calculator's screen. The pixels will cause discrepancies in the spacing of parts of the drawing, especially trees.

These instructions create an orchard of radius 6. First make any functions in the Y= editor inactive by deselecting the = sign for each function. Next, make the viewing window "square" by pressing WINDOW and adjusting the window variables to match the screen shown below. A "square" window has each pixel representing the same units in the *x*- and *y*-dimensions—so for this drawing, it means a circle will look like a circle, because it's not stretched in either direction.

```
WINDOW
 Xmin=-9.1
 Xmax=9.1
 Xscl=1
 Ymin=-6
 Ymax=6
 Yscl=.1
 Xres=1
```

Next, set the graph format to draw a grid; the grid dots will represent the trees. Press 2ND [FORMAT], and highlight settings as shown here.

```
RectGC PolarGC
CoordOn CoordOff
GridOff GridOn
AxesOn AxesOff
LabelOff LabelOn
ExprOn ExprOff
```

The final step is to draw the circular outline of the orchard. Go to the home screen (press 2ND [QUIT]). Press 2ND [DRAW], scroll down to **9:Circle(**, and press ENTER. Type **0,0,6** to indicate the center of the circle, (0, 0), and radius, 6. Press ENTER. You will see a circle outlining an orchard with radius 6.

```
Circle(0,0,6)
```

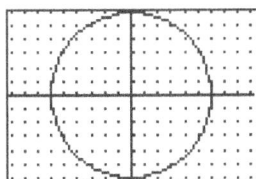

You might also draw a line segment through this picture to represent a line of sight. Press 2ND [DRAW], scroll down to **2:Line(**, and press ENTER. Use the arrow keys to move the cursor to the location of the first endpoint of the desired segment and press ENTER, then move to the second endpoint and press ENTER again.

You can also represent a line of sight in your drawing by entering a linear equation in the Y= editor. Once the linear equation has been entered, enter the command **Circle(0,0,6)**. The screen here represents a radius-4 orchard with the line **Y₁=0.3X** representing a line of sight.

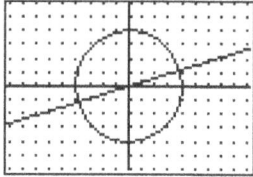

Storing and Recalling a Picture

To save a picture you've drawn on the graphing calculator, press 2ND [DRAW], move the cursor right to **STO**, and press ENTER to select **1:StorePic**. Next, press VARS, select **4:Picture...**, and select the memory location to store this picture. (You might choose the memory location according to the radius of the orchard, so an orchard with radius 4 is stored in **Pic4**.) The location you choose, such as **Pic1**, will appear on the home screen after the **StorePic** command, as shown here. Press ENTER one more time to carry out your **StorePic Pic1** command.

```
StorePic Pic1
```

To retrieve a stored picture, press 2ND [DRAW], move the cursor right to **STO**, select **2:RecallPic**, and press ENTER. Next, press VARS, select **4:Picture...**, and select the stored picture you'd like to recall. Press ENTER to select it, then ENTER again to carry out the command. The stored picture will be displayed.

```
RecallPic Pic1
```

Programming an Orchard Hideout

This program models orchards with various radii and tree sizes. For example, here is a radius-3 orchard with trees of radius 0.30 units.

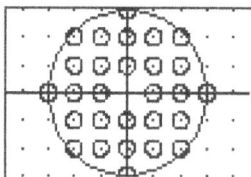

The program allows you to enter the radius of the orchard and the radius of the trees in the orchard. You will be asked to confirm if you enter a tree radius larger than half a unit. You also can select to have the lattice grid on or off in the final drawing.

Recall that the calculator is limited in its drawings (and graphs) by the location of the pixels. These drawings may not be useful for exact measurements, but they can serve to help explore varying orchard conditions.

To enter the program instructions, you must first be in the **PRGM** editor (if programming is new to you, refer to the Calculator Note "Programming the Calculator" in *Calculator Basics,* provided in the Year 3 general resources). Name your program "ORCHARDS," then enter the steps for the program from the column on the left. The column on the right explains each programming instruction. The text in italics summarizes the different parts of the program.

Instruction	Explanation
These first commands prepare the calculator's graphing window.	
:1→Xscl	
:1→Yscl	Press STO for the symbol →. Find **Xscl** by pressing VARS and selecting **1:Window...** and then **3:Xscl**. Find **Yscl** in the same way as **Xscl**. These two commands set the *x*- and *y*-scales to 1 for use with the calculator's grid command. (The program user will later have the option to place a dot at every lattice point.)
:GridOff	
:AxesOn	These two commands set up the graphing screen for drawing the orchard. Find them by pressing 2ND [FORMAT].
:PlotsOff	

:FnOff	These two commands eliminate plotting of data in lists and turn off any expressions in the function editor. Enter **PlotsOff** by pressing 2ND [STAT PLOT] and selecting **4:PlotsOff**. To enter **FnOff**, press VARS, highlight the **Y-VARS** menu, and select **On/Off** and then **2:FnOff**.
:ClrDraw	Clears any previous drawing. Find **ClrDraw** in the 2ND [DRAW] menu.

The next section contains commands that allow the user to input the radius of the orchard.

:ClrHome	Clears the home screen. Find **ClrHome** in the PRGM **I/O** menu.

:Disp "ORCHARD RADIUS?"

:Disp "(WORKS BEST WITH"

:Disp "VALUES UNDER 8)"

:Input R	This group of instructions asks the user to select the orchard radius and store the size as **R**. Find **Disp** and **Input** in the PRGM **I/O** menu. Press ALPHA to enter letters, or 2ND [A-LOCK] to lock in ALPHA entry mode. Press ALPHA [_] (above the 0 key) to enter a space between the words, and press ALPHA ["] (above the + key) for a quotation mark.
:abs(R)→R	This command ensures that the rest of the program uses a positive value for **R**. Find **abs(** in the MATH **NUM** menu.
:Lbl A	This label is a marker to return to if the user chooses to change the tree radius. Find **Lbl** in the PRGM **CTL** menu.

These commands allow the user to enter a tree radius. If the radius entered is larger than 0.5, the calculator will ask if the user wishes to change the radius.

:ClrHome

:Input "TREE RADIUS?",T

	Asks the user to input the tree radius, and stores that value in **T**.
:abs(T)→T	This command ensures that the program uses a positive value for **T**. Find **abs(** in the MATH **NUM** menu.
:If T>.5	Find **If** in the PRGM **CTL** menu. Find > in the 2ND [TEST] menu.
:Then	Find **Then** in the PRGM **CTL** menu.

:Disp "DO YOU REALLY"

:Disp "WANT TREES WITH"

:Disp "A RADIUS BIGGER"

:Disp "THAN .5 UNITS?"

:Input "YES(1)/NO(0)?",E	This set of instructions gives the program user a chance to reevaluate a tree radius choice that may cause trees to overlap. Use ÷ for "/".
:If E! 1	Find ≠ in the 2ND [TEST] menu.
:Goto A	If the user enters anything but **1** (yes), the program returns to **Lbl A** to ask for the tree radius again. Find **Goto** in the PRGM **CTL** menu.
:END	Ends the if-then statement. Press PRGM and scroll down for **END**.

This section asks whether the user would like to display a lattice grid.

:Menu("LATTICE GRID?","ON",B,"OFF",C)	This is a single, long line of code that the calculator will wrap around the screen. **Menu** is a command that allows users to select their choice from a screen menu. Depending on the user's selection, the program will go to **Lbl B** or **Lbl C**. Find **Menu** in the PRGM **CTL** menu.

:Lbl B

:GridOn	Find **GridOn** in the 2ND [FORMAT] screen.

This section contains instructions for the calculator to actually draw the orchard.

:Lbl C

:R→Ymax

:-R→Ymin

:1.516*R→Xmax

:-1.516*R→Xmin	These four commands set an approximately square window based on the radius of the orchard. "Square" means a circle will appear as a circle. To find **Ymax**, press VARS, select **1:Window**, and then select from the **X/Y** menu. Find the other commands similarly.
:Circle(0,0,R)	Draws a circle of radius **R** centered at (0,0). Find **Circle(** by pressing 2ND [DRAW] and scrolling down the **DRAW** menu.

This set of commands draws a tree at each of the lattice points inside the given orchard, excluding the point (0,0).

:For(X,-R,R)

:For(Y,-R,R)	The first of these **For** loops steps **X** from **−R** to **R**, that is, from left to right across the orchard. For each **X** value,

the second **For** loop then steps **Y** from **–R** to **R**. Together, the two loops step through every lattice point in the orchard's circumscribed circle. Find **For(** in the PRGM **CTL** menu.

:If X²+Y²" R² and (X! 0 or Y! 0)   This test runs every time through the loop to see if the (**X, Y**) coordinates fall inside the orchard but are not (0,0). Nothing will happen if either "**X²+Y²" R²**" or "**X! 0 or Y! 0**" is false. Find **or** and **and** by pressing 2ND [TEST] and highlighting **LOGIC**.

:Circle(X,Y,T)   If both **If** conditions are true, this instruction draws a circle (tree) of radius **T** with center (**X, Y**).

:End   Identifies the end of the inner (the **Y** value) **For** loop. Find **End** in the PRGM **CTL** menu.

:End   Identifies the end of the outer (the **X** value) **For** loop.

Using 2ND [ANS] and the Last Entry

You can recall the answer to the last calculation and use it to continue calculations. This may be more convenient than keying in a long decimal value from the previous answer, and more accurate than keying in only some digits of that answer.

Your calculator stores the last answer in **Ans**. You can use the variable **Ans** to represent the last answer in most cases. Place **Ans** in the cursor location by pressing 2ND [ANS] (above the (–) key).

You can use **Ans** as the first entry of the next calculation without having to press 2ND [ANS]. If you begin a new line with a function or operation key (such as +, –, ×, ÷, x^2, or ^), your calculator will automatically insert **Ans**.

You can use this last-answer recall to help keep track of all the multi-digit values while working through the activity *Orchard Growth Revisited*. One potential series of entries may look like the screen shown here.

```
2.5/(2π)
          .3978873577
12²*π-Ans²*π
           451.8919829
Ans/1.5
           301.2613219
```

An alternative method for using the multi-digit values is to store the result of a calculation into a variable before you evaluate another expression. To store a value, press STO> (this creates the → on screen), followed by a variable name. For example, you may want to remember the cross-sectional area of the original tree to use in later calculations. After calculating the cross-sectional area of the original tree (≈0.497 square inches), store the value in variable **A**. In the next step, after calculating the cross-sectional area of the tree with a trunk radius of 1 foot (≈452.39 square inches), you can find the required increase in cross-sectional area by evaluating **Ans–A**. Your calculator will display the result of this expression, **451.8919829**.

```
          .3978873577
Ans²*π
          .4973591972
Ans→A
          .4973591972
12²*π
          452.3893421
Ans–A
```

www.ingramcontent.com/pod-product-compliance
Lightning Source LLC
Chambersburg PA
CBHW051345200326
41521CB00014B/2485